CliffsNotes Differential Equations

Steven A. Leduc

Series Editor
Jerry Borow, Ph.D

CliffsQuickReview™ Differential Equations

Published by:

Learneo, Inc.

www.learneo.com

Note: If you purchased this book without a cover, you should be aware that this book is stolen property. It was reported as "unsold and destroyed" to the publisher, and neither the author nor the publisher has received any payment for this "stripped book."

Library of Congress Control Number:
ISBN: 978-1-957671-21-5

10 9 8 7 6 5
1B/TR/QR/QX/IN

CONTENTS

Before embarking on a study of differential equations, it is essential to solidify your technical facility with differentiation and integration. This section serves to review the definitions and techniques of differentiation and integration and then to introduce you to the study of differential equations.

Differentiation

Given a function $y = f(x)$, its **derivative**, denoted by y' or dy/dx, gives the instantaneous rate of change of f with respect to x. Geometrically, this gives the slope of the curve (that is, the slope of the tangent line to the curve) $y = f(x)$.

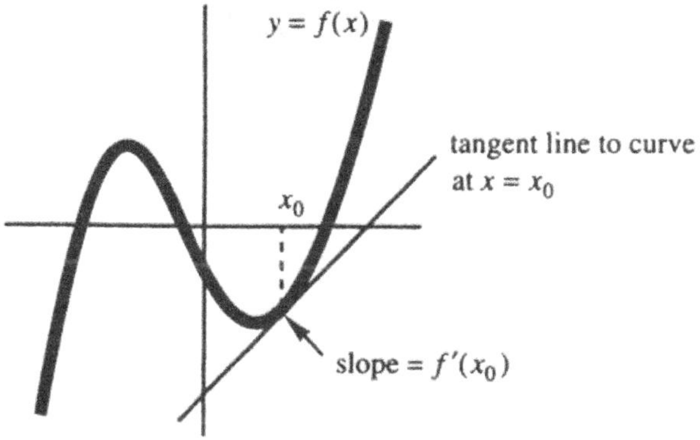

■ Figure 1 ■

The second derivative identifies the **concavity** of the curve $y = f(x)$. A portion of a differentiable curve $y = f(x)$ from $x = a$ to $x = b$ is said to be **concave up** if the curve lies above its tangent lines between a and b, and **concave down** if it lies below its tangent lines.

A curve $y = f(x)$ is concave up at those points x where the second derivative is positive, and concave down where the second derivative is negative. Points where the concavity changes are called **inflection points** and are located at those points x_0 where $f''(x_0) = 0$ but $f'''(x_0) \neq 0$.

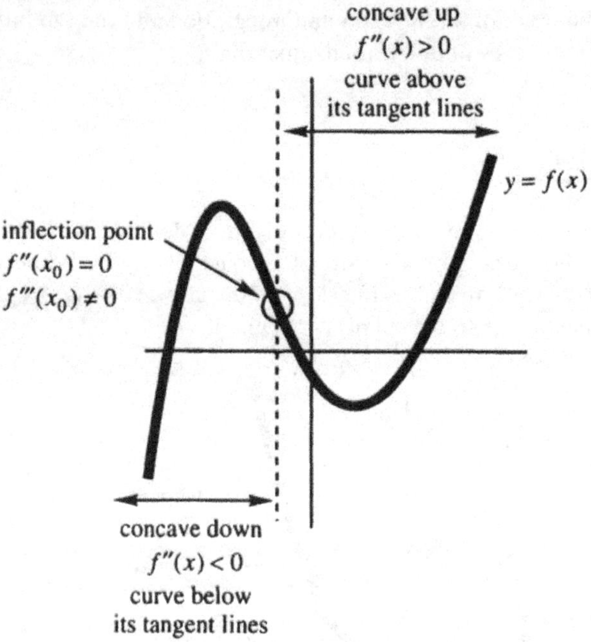

■ Figure 2 ■

Table 1 below lists the most frequently used properties of derivatives and, for later reference, the corresponding properties of integrals.

Table 1
COMPUTATIONAL PROPERTIES OF DIFFERENTIATION AND INTEGRATION

Differentiation	Integration
Linearity:	Linearity:
$d(ku) = k\,du$	$\int k\,du = k \int du = ku$
$d(u + v) = du + dv$	$\int (du + dv) = u + v$
Product rule:	Integration by parts:
$d(uv) = u\,dv + v\,du$	$\int u\,dv = uv - \int v\,du$
Quotient rule:	
$d\left(\frac{u}{v}\right) = \frac{v\,du - u\,dv}{v^2}$	
Chain rule:	Integration by substitution:
$(u \circ v)' = (u' \circ v) \cdot v'$	$\int (u' \circ v) \cdot v' = u \circ v$

In addition to being familiar with the definitions and fundamental properties, you should, of course, be able to actually differentiate a function. Although Table 2 below does not contain every differentiation formula, it will probably suffice to differentiate almost every function you are likely to encounter in practice. Again, for later reference, integration formulas are listed alongside the corresponding differentiation formulas. (Note: To avoid the repetition of writing "+ c" after every result in the right-hand column, the arbitrary additive constant c has been omitted from each of the integration formulas, as in Table 1 above.)

Table 2
DIFFERENTIATION AND INTEGRATION FORMULAS

Differentials	Integrals
$d(\text{constant}) = 0$	
$d(u^n) = nu^{n-1}\,du$	$\int u^n\,du = \frac{u^{n+1}}{n+1} \quad (n \neq -1)$
$d(e^u) = e^u\,du$	$\int e^u\,du = e^u$
$d(\ln u) = \frac{1}{u}\,du$	$\int \frac{1}{u}\,du = \ln\|u\|$
$d(\sin u) = \cos u\,du$	$\int \cos u\,du = \sin u$
$d(\cos u) = -\sin u\,du$	$\int \sin u\,du = -\cos u$
$d(\tan u) = \sec^2 u\,du$	$\int \sec^2 u\,du = \tan u$
$d(\cot u) = -\csc^2 u\,du$	$\int \csc^2 u\,du = -\cot u$
$d(\sec u) = \sec u \tan u\,du$	$\int \sec u\,du = \ln\|\sec u + \tan u\|$
$d(\csc u) = -\csc u \cot u\,du$	$\int \csc u\,du = -\ln\|\csc u + \cot u\|$
$d(\arcsin u) = \frac{1}{\sqrt{1-u^2}}\,du$	$\int \frac{1}{\sqrt{1-u^2}}\,du = \arcsin u$
$d(\arctan u) = \frac{1}{1+u^2}\,du$	$\int \frac{1}{1+u^2}\,du = \arctan u$

Example 1: Differentiate each of the following:

(a) $y = 3x^2 - 5x + 8$

(b) $y = x^2 e^x$

(c) $y = \ln x / x$

(d) $y = (x^3 + x - 1)^4$

(e) $y = \sqrt{x^2 + 1}$

(f) $y = \sin(x^2)$

(g) $y = \sin^2 x$

(h) $y = e^{\tan x}$

(i) $y = \csc(\sin \sqrt{x})$

The solutions are as follows:

(a) $y' = 6x - 5$

(b) Using the product rule, $y' = x^2 \cdot e^x + e^x \cdot 2x = xe^x(x + 2)$

(c) By the quotient rule,

$$y' = \frac{x \cdot \frac{1}{x} - \ln x \cdot 1}{x^2} = \frac{1 - \ln x}{x^2}$$

All of the remaining parts use the chain rule (as embodied in the formulas in Table 2).

(d) $y' = 4(x^3 + x - 1)^3 \cdot (3x^2 + 1)$

(e) $y = (x^2 + 1)^{1/2} \Rightarrow y' = \frac{1}{2}(x^2 + 1)^{-1/2} \cdot 2x = \dfrac{x}{\sqrt{x^2 + 1}}$

(f) $y' = 2x\cos(x^2)$

(g) $y = (\sin x)^2 \Rightarrow y' = 2\sin x \cos x = \sin 2x$

(h) $y' = e^{\tan x}\sec^2 x$

(i) $y' = -\csc(\sin\sqrt{x})\cot(\sin\sqrt{x}) \cdot \dfrac{\cos\sqrt{x}}{2\sqrt{x}}$ ■

Example 2: What is the equation of the tangent line to the curve $y = e^x \ln x$ at the point (1, 0)?

The first step is to find the slope of the tangent line at $x = 1$, which is the value of the derivative of y at this point:

$$\text{slope at point } (1, 0) = \left.\frac{dy}{dx}\right|_{x=1} = \left[e^x \cdot \frac{1}{x} + \ln x \cdot e^x\right]_{x=1} = e$$

Since the **point-slope formula** says that the straight line with slope m which passes through the point (x_0, y_0) has the equation

$$y - y_0 = m(x - x_0)$$

the equation of the desired tangent line is $y = e(x - 1)$. ■

Example 3: Is the curve $y = \arcsin\sqrt{x}$ concave up or is it concave down at the point $(\frac{1}{4}, \frac{\pi}{6})$?

Concavity is determined by the sign of the second derivative.

Since

$$d(\arcsin u) = \frac{1}{\sqrt{1 - u^2}}\,du$$

the first derivative of $y = \arcsin \sqrt{x}$ is

$$y' = \frac{1}{\sqrt{1 - x}} \cdot \frac{1}{2\sqrt{x}} = \frac{1}{2}(x - x^2)^{-1/2}$$

Its second derivative is therefore

$$y'' = -\frac{1}{4}(x - x^2)^{-3/2}(1 - 2x) = \frac{2x - 1}{4\sqrt{(x - x^2)^3}}$$

For $x = \frac{1}{4}$, the denominator in the expression above for y'' is positive (as it is for any x in the interval $0 < x < 1$), but the numerator is negative. Therefore, $y''(\frac{1}{4}) < 0$, and the curve is concave down at the point $(\frac{1}{4}, \frac{\pi}{6})$. ■

Example 4: Consider the curve given implicitly by the equation

$$3x^2y - y^3 = x + 1$$

What is the slope of this curve at the point where it crosses the x axis?

To find the slope of a curve defined implicitly (as is the case here), the technique of **implicit differentiation** is used: Differentiate both sides of the equation with respect to x; then solve the resulting equation for y'.

$$\begin{aligned} 3x^2y - y^3 &= x + 1 \\ (3x^2y' + 6xy) - 3y^2y' &= 1 \\ y'(3x^2 - 3y^2) &= 1 - 6xy \\ y' &= \frac{1 - 6xy}{3(x^2 - y^2)} \end{aligned}$$

The curve crosses the x axis when $y = 0$, and the given equation clearly implies that $x = -1$ at $y = 0$. From the expression directly above, the slope of the curve at the point $(-1, 0)$ is

$$y'\big|_{(-1,0)} = \frac{1-6xy}{3(x^2-y^2)}\bigg|_{(-1,0)} = \frac{1}{3} \quad \blacksquare$$

Partial Differentiation

Given a function of two variables, $f(x, y)$, the derivative with respect to x only (treating y as a constant) is called the **partial derivative of f with respect to x** and is denoted by either $\partial f/\partial x$ or f_x. Similarly, the derivative of f with respect to y only (treating x as a constant) is called the **partial derivative of f with respect to y** and is denoted by either $\partial f/\partial y$ or f_y.

The **second partial derivatives** of f come in four types:

	Notation
■ Differentiate f with respect to x twice. (That is, differentiate f with respect to x; then differentiate the result with respect to x again.)	$\frac{\partial^2 f}{\partial x^2}$ or f_{xx}
■ Differentiate f with respect to y twice. (That is, differentiate f with respect to y; then differentiate the result with respect to y again.)	$\frac{\partial^2 f}{\partial y^2}$ or f_{yy}

Mixed partials:

■ First differentiate f with respect to x; then differentiate the result with respect to y.	$\frac{\partial^2 f}{\partial y \partial x}$ or f_{xy}

- First differentiate f with respect to y; then differentiate the result with respect to x. $\dfrac{\partial^2 f}{\partial x \partial y}$ or f_{yx}

For virtually all functions $f(x, y)$ commonly encountered in practice, f_{yx} will be identical to f_{xy}; that is, the order in which the derivatives are taken in the mixed partials is immaterial.

Example 5: If $f(x, y) = 3x^2y + 5x - 2y^2 + 1$, find $f_x, f_y, f_{xx}, f_{yy}, f_{xy}$, and f_{yx}.

First, differentiating f with respect to x (while treating y as a constant) yields

$$f_x = 6xy + 5$$

Next, differentiating f with respect to y (while treating x as a constant) yields

$$f_y = 3x^2 - 4y$$

The second partial derivative f_{xx} means the partial derivative of f_x with respect to x; therefore,

$$f_{xx} = (f_x)_x = \frac{\partial}{\partial x}(f_x) = \frac{\partial}{\partial x}(6xy + 5) = 6y$$

The second partial derivative f_{yy} means the partial derivative of f_y with respect to y; therefore,

$$f_{yy} = (f_y)_y = \frac{\partial}{\partial y}(f_y) = \frac{\partial}{\partial y}(3x^2 - 4y) = -4$$

The mixed partial f_{xy} means the partial derivative of f_x with respect to y; therefore,

$$f_{xy} = (f_x)_y = \frac{\partial}{\partial y}(f_x) = \frac{\partial}{\partial y}(6xy + 5) = 6x$$

The mixed partial f_{yx} means the partial derivative of f_y with respect to x; therefore,

$$f_{yx} = (f_y)_x = \frac{\partial}{\partial x}(f_y) = \frac{\partial}{\partial x}(3x^2 - 4y) = 6x$$

Note that $f_{yx} = f_{xy}$, as expected. ■

Integration

Indefinite integration means *antidifferentiation*; that is, given a function $f(x)$, determine the most general function $F(x)$ whose derivative is $f(x)$. The symbol for this operation is the integral sign, $\int$, followed by the **integrand** (the function to be integrated) and a differential, such as dx, which specifies the variable of integration.

On the other hand, the fundamental geometric interpretation of **definite integration** is to *compute an area*. That is, given a function $f(x)$ and an interval $a \le x \le b$ in its domain, the definite integral of f from a to b gives the area bounded by the curve $y = f(x)$, the x axis, and the vertical lines $x = a$ and $x = b$. The symbol for this operation is the integral sign with **limits of integration** (a and b), $\int_a^b$, followed by the function and the differential which specifies the variable of integration.

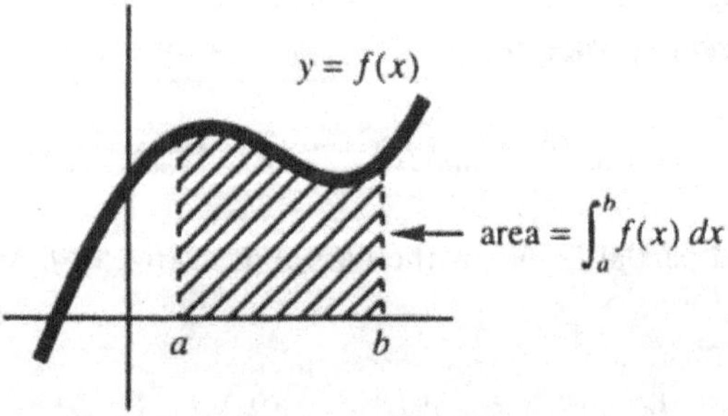

■ Figure 3 ■

From their definitions, you can see that the processes of indefinite integration and definite integration are really very different. The *indefinite* integral of a function is the collection of functions which are its antiderivatives, whereas the *definite* integral of a function requires two limits of integration and gives a numerical result equal to an area in the *xy* plane. However, the fact that both operations are called "integration" and are denoted by such similar symbols suggests that there is a link between them.

The **Fundamental Theorem of Calculus** says that differentiation (finding the slope of a curve) is the inverse operation of definite integration (finding the area under a curve). More explicitly, Part I of the Fundamental Theorem says that if a function is integrated (to form a definite integral with a variable upper limit of integration), and the result is then differentiated, the original function is recovered; that is, differentiation "undoes" integration. Part II gives the connection between definite and indefinite integrals. It says that a definite integral can be computed by first determining an indefinite integral (so computing the area under a curve is done by antidifferentiating).

The Fundamental Theorem of Calculus (Part I):

If f is continuous, then $d\ dx \int_a^x f(t)\,dt = f(x)$.

The Fundamental Theorem of Calculus (Part II):

If f is continuous with antiderivative F, then $\int_a^b f(x)\,dx = F(b) - F(a)$.

Example 6: Evaluate the integral

$$\int (x^4 - 3x^2 + x - 1)\, dx$$

Using the first integration formula in Table 2, every function whose derivative equals $f(x) = x^4 - 3x^2 + x - 1$ is given by

$$\int (x^4 - 3x^2 + x - 1)\, dx = \tfrac{1}{5}x^5 - x^3 + \tfrac{1}{2}x^2 - x + c$$

where c is an arbitrary constant. ■

Techniques of Indefinite Integration

Integration by substitution. This section opens with integration **by substitution,** the most widely used integration technique, illustrated by several examples. The idea is simple: Simplify an integral by letting a single symbol (say the letter u) stand for some complicated expression in the integrand. If the differential of u is left over in the integrand, the process will be a success.

Example 7: Determine

$$\int x\sqrt{x^2 + 1}\, dx$$

Let $u = x^2 + 1$ (this is the substitution); then $du = 2x\, dx$, and the given integral is transformed into

$$\int \sqrt{x^2 + 1} \cdot x\, dx = \int \sqrt{u} \cdot \tfrac{1}{2}\, du = \tfrac{1}{2} \int u^{1/2}\, du = \tfrac{1}{2} \cdot \tfrac{2}{3} u^{3/2} + c$$

which transforms back to $\frac{1}{3}(x^2 + 1)^{3/2} + c$. ■

Example 8: Integrate

$$\int \sin^3 x \cos x \, dx$$

Let $u = \sin x$; then $du = \cos x \, dx$, and the given integral becomes

$$\int u^3 \, du = \tfrac{1}{4}u^4 + c = \tfrac{1}{4}\sin^4 x + c \qquad \blacksquare$$

Example 9: Evaluate

$$\int \tan x \, dx$$

First, rewrite $\tan x$ as $\sin x / \cos x$; then let $u = \cos x$, $du = -\sin x \, dx$:

$$\int \tan x \, dx = \int \frac{\sin x}{\cos x} \, dx = \int \frac{-du}{u} = -\ln|u| + c = -\ln|\cos x| + c \quad \blacksquare$$

Example 10: Evaluate

$$\int xe^{x^2} \, dx$$

Let $u = x^2$; then $du = 2x \, dx$, and the integral is transformed into

$$\int xe^{x^2} \, dx = \int e^u \cdot \tfrac{1}{2} \, du = \tfrac{1}{2}e^u + c = \tfrac{1}{2}e^{x^2} + c \qquad \blacksquare$$

Example 11: Determine

$$\int \sec^2 x \tan x \, dx$$

Let $u = \sec x$; then $du = \sec x \tan x \, dx$, and the integral is transformed into

$$\int \sec^2 x \tan x \, dx = \int \sec x \cdot \sec x \tan x \, dx = \int u \, du = \tfrac{1}{2}u^2 + c$$

$$= \tfrac{1}{2}\sec^2 x + c \qquad \blacksquare$$

Integration by parts. The product rule for differentiation says $d(uv) = u\,dv + v\,du$. Integrating both sides of this equation gives $uv = \int u\,dv + \int v\,du$, or equivalently

$$\int u\,dv = uv - \int v\,du$$

This is the formula for **integration by parts**. It is used to evaluate integrals whose integrand is the product of one function (u) and the differential of another (dv). Several examples follow.

Example 12: Integrate

$$\int xe^x\,dx$$

Compare this problem with Example 10. A simple substitution made that integral trivial; unfortunately, such a simple substitution would be useless here. This is a prime candidate for integration by parts, since the integrand is the product of a function (x) and the differential ($e^x\,dx$) of another, and when the formula for integration by parts is used, the integral that is left is easier to evaluate (or, in general, at least not more difficult to integrate) than the original.

Let $u = x$ and $dv = e^x\,dx$; then

$$u = x \qquad v = e^x$$

$$du = dx \qquad dv = e^x\,dx$$

and the formula for integration by parts yields

$$\int u\,dv = uv - \int v\,du$$

$$\int xe^x\,dx = xe^x - \int e^x\,dx$$

$$= xe^x - e^x + c$$

$$= e^x(x - 1) + c \qquad \blacksquare$$

Example 13: Integrate

$$\int x \cos x \, dx$$

Let $u = x$ and $dv = \cos x \, dx$; then

$$u = x \qquad v = \sin x$$

$$du = dx \qquad dv = \cos x \, dx$$

The formula for integration by parts gives

$$\int u \, dv = uv - \int v \, du$$

$$\int x \cos x \, dx = x \sin x - \int \sin x \, dx$$

$$= x \sin x + \cos x + c \qquad \blacksquare$$

Example 14: Evaluate

$$\int \ln x \, dx$$

Let $u = \ln x$ and $dv = dx$; then

$$u = \ln x \qquad v = x$$

$$du = \frac{1}{x} dx \qquad dv = dx$$

and the formula for integration by parts yields

$$\int u \, dv = uv - \int v \, du$$

$$\int \ln x \, dx = x \ln x - \int x \cdot \frac{1}{x} dx$$

$$= x \ln x - x + c \qquad \blacksquare$$

Partial Integration

Suppose it is known that a given function $f(x)$ is the derivative of some function $F(x)$; how is $F(x)$ found? The answer, of course, is to integrate $f(x)$. Now consider a related question: Suppose it is known that a given function $f(x, y)$ is the *partial* derivative with respect to x of some function $F(x, y)$; how is $F(x, y)$ found? The answer is to integrate $f(x, y)$ with respect to x, a process I refer to as **partial integration**. Similarly, suppose it is known that a given function $f(x, y)$ is the partial derivative with respect to y of some function $F(x, y)$; how is $F(x, y)$ found? Integrate $f(x, y)$ with respect to y.

Example 15: Let $M(x, y) = 2xy^2 + x^2 - y$. It is known that M equals f_x for some function $f(x, y)$. Determine the most general such function $f(x, y)$.

Since $M(x, y)$ is the partial derivative with respect to x of some function $f(x, y)$, M must be partially integrated with respect to x to recover f. This situation can be symbolized as follows:

$$f(x, y) \xrightarrow{\frac{\partial}{\partial x}} M(x, y)$$

$$f(x, y) \xleftarrow[\int(\cdot)\partial x]{} M(x, y)$$

Therefore,

$$\begin{aligned} f(x, y) &= \int M(x, y)\partial x \\ &= \int (2xy^2 + x^2 - y)\partial x \end{aligned}$$

$$f(x, y) = x^2y^2 + \tfrac{1}{3}x^3 - xy + \psi(y)$$

Note carefully that the "constant" of integration here is any (differentiable) function of y—denoted by $\psi(y)$—since any such function would vanish upon partial differentiation with respect to x (just as any pure constant c would vanish upon ordinary differentiation). If the question had asked merely for *a* function $f(x, y)$ for which $f_x = M$, you could just take $\psi(y) \equiv 0$. ■

Example 16: Let $N(x, y) = \sin x \cos y - xy + 1$. It is known that N equals f_y for some function $f(x, y)$. Determine the most general such function $f(x, y)$.

Since $N(x, y)$ is the partial derivative with respect to y of some function $f(x, y)$, N must be partially integrated with respect to y to recover f. This situation can be symbolized as follows:

$$f(x, y) \xrightarrow{\frac{\partial}{\partial y}} N(x, y)$$
$$f(x, y) \xleftarrow[\int(\cdot)\partial y]{} N(x, y)$$

Therefore,

$$\begin{aligned} f(x, y) &= \int N(x, y)\partial y \\ &= \int (\sin x \cos y - xy + 1)\partial y \\ f(x, y) &= \sin x \sin y - \tfrac{1}{2}xy^2 + y + \xi(x) \end{aligned}$$

Note carefully that the "constant" of integration here is any (differentiable) function of x—denoted by $\xi(x)$—since any such function would vanish upon partial differentiation with respect to y. If the question had asked merely for *a* function $f(x, y)$ for which $f_y = N$, you could just take $\xi(x) \equiv 0$. ■

Introduction to Differential Equations

In high school, you studied algebraic equations like

$$3x - 2(x - 4) = 5x, \quad x^2 - 8x + 15 = 0, \quad \text{and} \quad |6x - 2| = 4$$

The goal here was to **solve the equation**, which meant to find the value (or values) of the variable that makes the equation true. For example, $x = 2$ is the solution to the first equation because only when 2 is substituted for the variable x does the equation become an identity (both sides of the equation are identical when and only when $x = 2$).

In general, each type of algebraic equation had its own particular method of solution; quadratic equations were solved by one method, equations involving absolute values by another, and so on. In each case, an equation was presented (or arose from a word problem), and a certain method was employed to arrive at a solution, a method appropriate for the particular equation at hand.

These same general ideas carry over to **differential equations**, which are equations involving derivatives. There are different types of differential equations, and each type requires its own particular solution method. The simplest differential equations are those of the form $y' = f(x)$. For example, consider the differential equation

$$\frac{dy}{dx} = 2x$$

It says that the derivative of some function y is equal to $2x$. To **solve the equation** means to determine the unknown (the function y) which will turn the equation into an identity upon substitution. In this case all that is needed to solve the equation is an integration:

$$dy = 2x\,dx$$

$$\int dy = \int 2x\,dx$$

$$y = x^2 + c$$

Thus, the **general solution** of the differential equation $y' = 2x$ is $y = x^2 + c$, where c is any arbitrary constant. Note that there are actually infinitely many **particular** solutions, such as $y = x^2 + 1$, $y = x^2 - 7$, or $y = x^2 + \pi$, since any constant c may be chosen.

Geometrically, the differential equation $y' = 2x$ says that at each point (x, y) on some curve $y = y(x)$, the slope is equal to $2x$. The solution obtained for the differential equation shows that this property is satisfied by any member of the **family** of curves $y = x^2 + c$ (any only by such curves); see Figure 4.

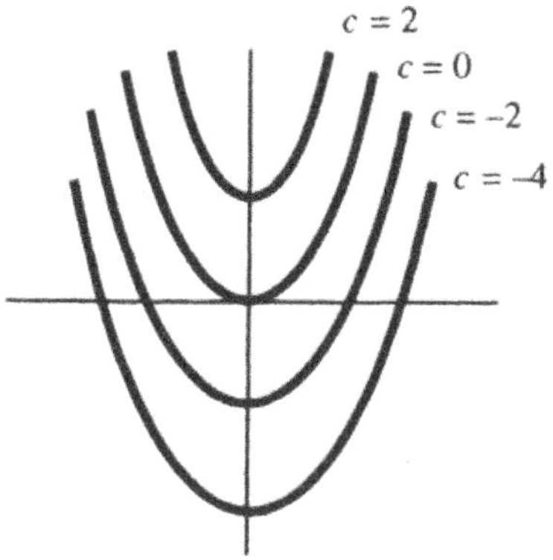

■ Figure 4 ■

Since these curves were obtained by solving a differential equation—which either explicitly or implicitly involves taking an integral—they are sometimes referred to as **integral curves** of the differential equation (particularly when these solutions are graphed). If one particular solution or integral curve is desired, the differential equation is appended with one or more supplementary conditions. These additional conditions uniquely specify the value of the arbitrary constant or constants in the general solution. For example, consider the problem

$$\frac{dy}{dx} = 2x \quad \text{and} \quad y = 2 \text{ when } x = 0$$

The **initial condition** "$y = 2$ when $x = 0$" is usually abbreviated "$y(0) = 2$," which is read "y at 0 equals 2." The combination of a differential equation and an initial condition (also known as a **constraint**) is called an **initial value problem** (abbreviated **IVP**).

For differential equations involving higher derivatives, two or more constraints may be present. If all constraints are given at the same value of the independent variable, then the term IVP still applies. If, however, the constraints are given at different values of the independent variable, the term **boundary value problem (BVP)** is used instead. For example,

this is an IVP: $y'' + 2y' - 3y = 0, \quad y(0) = 1, \quad y'(0) = 5$

(the 0 in $y(0)$ and the 0 in $y'(0)$: same)

but

this is a BVP: $y'' + 2y' - 3y = 0, \quad y(0) = 1, \quad y'(1) = 5$

(the 0 in $y(0)$ and the 1 in $y'(1)$: different)

To solve an IVP or BVP, first find the general solution of the differential equation and then determine the value(s) of the arbitrary constant(s) from the constraints.

Example 17: Solve the IVP

$$y' = 2x$$

$$y(0) = 2$$

As discussed above, the general solution of this differential equation is the family $y = x^2 + c$. Since the constraint says that y must equal 2 when x is 0,

$$y(0) = 2 \Rightarrow [x^2 + c]_{x=0} = 2 \Rightarrow 0^2 + c = 2 \Rightarrow c = 2$$

so the solution of this IVP is $y = x^2 + 2$. ■

Example 18: Consider the differential equation $y'' + 2y' - 3y = 0$. Verify that $y = c_1e^x + c_2e^{-3x}$ (where c_1 and c_2 are arbitrary constants) is a solution. Given that *every* solution of this differential equation can be written in the form $y = c_1e^x + c_2e^{-3x}$, solve the IVP

$$y'' + 2y' - 3y = 0$$
$$y(0) = 1$$
$$y'(0) = 5$$

To verify that $y = c_1e^x + c_2e^{-3x}$ is a solution of the differential equation, substitute. Since

$$y' = c_1e^x - 3c_2e^{-3x} \quad \text{and} \quad y'' = c_1e^x + 9c_2e^{-3x}$$

once $c_1e^x + c_2e^{-3x}$ is substituted for y, the left-hand side of the differential equation becomes

$$\begin{aligned} y'' + 2y' - 3y &= (c_1e^x + 9c_2e^{-3x}) + 2(c_1e^x - 3c_2e^{-3x}) \\ &\quad - 3(c_1e^x + c_2e^{-3x}) \\ &= (c_1e^x + 2c_1e^x - 3c_1e^x) \\ &\quad + (9c_2e^{-3x} - 6c_2e^{-3x} - 3c_2e^{-3x}) \\ &= 0 \quad \checkmark \end{aligned}$$

Now, to satisfy the conditions $y(0) = 1$ and $y'(0) = 5$, the constants c_1 and c_2 must be chosen so that

$$y(0) = 1 \Rightarrow [c_1e^x + c_2e^{-3x}]_{x=0} = 1 \Rightarrow c_1 + c_2 = 1$$

and

$$y'(0) = 5 \Rightarrow [c_1e^x - 3c_2e^{-3x}]_{x=0} = 5 \Rightarrow c_1 - 3c_2 = 5$$

Solving these two equations yields $c_1 = 2$ and $c_2 = -1$. Thus, the particular solution specified by the given IVP is $y = 2e^x - e^{-3x}$. ■

The **order** of a differential equation is the order of the highest derivative that appears in the equation. For example, $y' = 2x$ is a first-order equation, $y'' + 2y' - 3y = 0$ is a second-order equation, and $y''' - 7y' + 6y = 12$ is a third-order equation. Note that the general solution of the first-order equation from Example 17 contained one arbitrary constant, and the general solution of the second-order equation in Example 18 contained two arbitrary constants. This phenomenon is not coincidental. In *most* cases, *the number of arbitrary constants in the general solution of a differential equation is the same as the order of the equation.*

Example 19: Solve the second-order differential equation $y'' = x + \cos x$.

Integrating both sides of the equation will yield a differential equation for y':

$$\begin{aligned} y' &= \int y'' \\ &= \int (x + \cos x)\, dx \\ &= \tfrac{1}{2}x^2 + \sin x + c_1 \end{aligned}$$

Integrating once more will give y:

$$\begin{aligned} y &= \int y' \\ &= \int (\tfrac{1}{2}x^2 + \sin x + c_1)\, dx \\ &= \tfrac{1}{6}x^3 - \cos x + c_1 x + c_2 \end{aligned}$$

where c_1 and c_2 are arbitrary constants. Note that there are two arbitrary constants in the general solution, which you should typically expect for a second-order equation. ■

Example 20: For the following IVP, find the solution valid for $x > 0$:

$$y''' = 105\sqrt{x} - \frac{2}{x^3} + 6$$

$$y(1) = 7$$

$$y'(1) = 37$$

$$y''(1) = 73$$

The general solution of a third-order differential equation typically contains three arbitrary constants, so an IVP involving a third-order differential equation will necessarily have three constraint equations (as is the case here). As in Examples 17 and 19, the given differential equation is of the form

$$y^{(n)} = f(x)$$

where $y^{(n)}$ denotes the nth derivative of the function y. These differential equations are the easiest to solve, since all they require are n successive integrations. Note how the first-order differential equation in Example 17 was solved with one integration, and the second-order equation in Example 19 was solved with two integrations. The third-order differential equation given here will be solved with three successive integrations. Here's the first:

$$y'' = \int (105x^{1/2} - 2x^{-3} + 6)\,dx$$

$$= 70x^{3/2} + x^{-2} + 6x + c_1$$

The value of this first arbitrary constant (c_1) can be found by applying the condition $y''(1) = 73$:

$$y''(1) = 73 \Rightarrow [70x^{3/2} + x^{-2} + 6x + c_1]_{x=1} = 73$$

$$\Rightarrow 70 + 1 + 6 + c_1 = 73 \Rightarrow c_1 = -4$$

Thus, $y'' = 70x^{3/2} + x^{-2} + 6x - 4$.

Now, perform the second integration, which will yield y':

$$y' = \int (70x^{3/2} + x^{-2} + 6x - 4)\,dx$$
$$= 28x^{5/2} - x^{-1} + 3x^2 - 4x + c_2$$

The value of this arbitrary constant (c_2) can be found by applying the constraint $y'(1) = 37$:

$$y'(1) = 37 \Rightarrow [28x^{5/2} - x^{-1} + 3x^2 - 4x + c_2]_{x=1} = 37$$
$$\Rightarrow 28 - 1 + 3 - 4 + c_2 = 37$$
$$\Rightarrow c_2 = 11$$

Therefore, $y' = 28x^{5/2} - x^{-1} + 3x^2 - 4x + 11$. Integrating once more will give the solution y:

$$y = \int (28x^{5/2} - x^{-1} + 3x^2 - 4x + 11)\,dx$$
$$= 8x^{7/2} - \ln x + x^3 - 2x^2 + 11x + c_3$$

The value of this arbitrary constant (c_3) can be found by applying the condition $y(1) = 7$:

$$y(1) = 7 \Rightarrow [8x^{7/2} - \ln x + x^3 - 2x^2 + 11x + c_3]_{x=1} = 7$$
$$\Rightarrow 8 - 0 + 1 - 2 + 11 + c_3 = 7$$
$$\Rightarrow c_3 = -11$$

Thus, the solution is $y = 8x^{7/2} - \ln x + x^3 - 2x^2 + 11x - 11$.

A few technical notes about this example:

- The given differential equation makes sense only for $x > 0$ (note the $\sqrt{x}$ and $2/x^3$ terms). To respect this restriction, the problem states the **domain** of the equation and its solution [that is, the set of values of the variable(s) where the equation and solution are valid] as $x > 0$. Always be aware of the domain of the solution.

- Although the integral of x^{-1} is usually written $\ln|x|$, the absolute value sign is not needed here, since the domain of the solution is $x > 0$, and $|x| = x$ for any $x > 0$.

- Contrast the methods used to evaluate the arbitrary constants in Examples 18 and 20. In Example 18, the constraints were applied all at once at the end. In Example 20, however, the constants were evaluated one at a time as the solution progressed. Both methods are valid, and each particular problem (and your preference) will suggest which to use. ■

Example 21: Find the differential equation for the family of curves $x^2 + y^2 = c^2$ (in the xy plane), where c is an arbitrary constant.

This problem is a reversal of sorts. Typically, you're given a differential equation and asked to find its family of solutions. Here, on the other hand, the general solution is given, and an expression for its defining differential equation is desired. Differentiating both sides of the equation (with respect to x) gives

$$x^2 + y^2 = c^2$$

$$2x + 2y\frac{dy}{dx} = 0$$

$$x + y\frac{dy}{dx} = 0$$

$$\frac{dy}{dx} = -\frac{x}{y}$$

This differential equation can also be expressed in another form, one that will arise quite often. By "cross multiplying," the differential equation directly above becomes

$$y\,dy = -x\,dx$$

which is then normally written with both differentials (the *dx* and the *dy*) together on one side:

$$x\,dx + y\,dy = 0$$

Either $y' = -x/y$ or $x\,dx + y\,dy = 0$ would be an acceptable way of writing the differential equation that defines the given family (of circles) $x^2 + y^2 = c^2$. ■

Example 22: Verify that the equation $y = \ln(x/y)$ is an implicit solution of the IVP

$$y\,dx - x(y + 1)\,dy = 0$$
$$y(e) = 1$$

First note that it is not always possible to express a solution in the form "y = some function of x." Sometimes when a differential equation is solved, the solution is most naturally expressed with y's (the dependent variable) on *both* sides of the equation, as in $y = \ln(x/y)$. Such a solution is called an **implicit** solution, as opposed to an **explicit** solution, which has y all by itself on one side of the equation and a function of x only on the right (as in $y = x^2 + 2$, for example). Implicit solutions are perfectly acceptable (in some cases, necessary) as long as the equation actually defines y as a function of x (even if an explicit formula for this function is not or cannot be found). However, explicit solutions are preferable when available.

Perhaps the simplest way to verify this implicit solution is to follow the procedure of Example 21: Find the differential equation for the solution $y = \ln(x/y)$. To simplify the work, first rewrite $\ln(x/y)$ as $\ln x - \ln y$:

$$y = \ln x - \ln y$$

$$\frac{dy}{dx} = \frac{1}{x} - \frac{1}{y}\frac{dy}{dx}$$

$$\frac{dy}{dx}\left(1+\frac{1}{y}\right)=\frac{1}{x}$$

$$\frac{dy}{dx}=\frac{1/x}{1+(1/y)}$$

$$\frac{dy}{dx}=\frac{1/x}{1+(1/y)}\cdot\frac{xy}{xy}$$

$$\frac{dy}{dx}=\frac{y}{x(y+1)}$$

$$y\,dx-x(y+1)\,dy=0$$

Therefore, the differential equation given in the statement of the problem is indeed correct. The initial condition is also satisfied, since $1=\ln(e/1)$ implies $y(e)=1$ satisfies $y=\ln(x/y)$. ■

Example 23: Discuss the solution to each of the differential equations

$$(y')^2+x^2=0 \quad \text{and} \quad (y')^2+y^2=0$$

The first differential equation has no solution, since no real-valued function $y=y(x)$ can satisfy $(y')^2=-x^2$ (because squares of real-valued functions can't be negative).

The second differential equation states that the sum of two squares is equal to 0, so both y' and y must be identically 0. This equation does have a solution, but it is only the constant function $y\equiv 0$. Note that this differential equation illustrates an exception to the general rule stating that the number of arbitrary constants in the general solution of a differential equation is the same as the order of the equation. Although $(y')^2+y^2$ is a first-order equation, its general solution $y\equiv 0$ contains no arbitrary constants at all. ■

One final note: Since there are two major categories of derivatives, **ordinary** derivatives like

$$\frac{dy}{dx} \equiv y' \quad \text{and} \quad \frac{d^2y}{dx^2} \equiv y''$$

and **partial** derivatives such as

$$\frac{\partial f}{\partial x} \equiv f_x \quad \text{and} \quad \frac{\partial^2 f}{\partial y \partial x} \equiv f_{xy}$$

there are two major categories of differential equations. **Ordinary differential equations (ODEs)**, like the equations in all of the examples above, involve ordinary derivatives, while **partial differential equations (PDEs)**, such as

$$\frac{\partial f}{\partial t} = k\frac{\partial^2 f}{\partial x^2} \quad \text{and} \quad \frac{\partial^2 w}{\partial x^2} + \frac{\partial^2 w}{\partial y^2} = 0$$

involve partial derivatives. Only ordinary differential equations are examined in this book.

From the Introduction, you know that a first-order differential equation is one containing a first—but no higher—derivative of the unknown function. For virtually every such equation encountered in practice, the general solution will contain one arbitrary constant, that is, one parameter, so a first-order IVP will contain one initial condition. There is no general method that solves every first-order equation, but there are methods to solve particular types.

Exact Equations

Given a function $f(x, y)$ of two variables, its **total differential** df is defined by the equation

$$df = \frac{\partial f}{\partial x}\,dx + \frac{\partial f}{\partial y}\,dy$$

Example 1: If $f(x, y) = x^2y + 6x - y^3$, then

$$df = (2xy + 6)\,dx + (x^2 - 3y^2)\,dy \quad \blacksquare$$

The equation $f(x, y) = c$ gives the family of integral curves (that is, the solutions) of the differential equation

$$df = 0$$

Therefore, if a differential equation has the form

$$\frac{\partial f}{\partial x}\,dx + \frac{\partial f}{\partial y}\,dy = 0 \qquad (*)$$

for some function $f(x, y)$, then it is automatically of the form $df = 0$.

so the general solution is immediately given by $f(x, y) = c$. In this case,

$$\frac{\partial f}{\partial x}\,dx + \frac{\partial f}{\partial y}\,dy$$

is called an **exact differential,** and the differential equation (*) is called an **exact equation.** To determine whether a given differential equation

$$M(x, y)\,dx + N(x, y)\,dy = 0$$

is exact, use the

Test for Exactness: A differential equation $M\,dx + N\,dy = 0$ is exact if and only if

$$\frac{\partial M}{\partial y} = \frac{\partial N}{\partial x}$$

Example 2: Is the following differential equation exact?

$$(y^2 - 2x)\,dx + (2xy + 1)\,dy = 0$$

The function that multiplies the differential dx is denoted $M(x, y)$, so $M(x, y) = y^2 - 2x$; the function that multiplies the differential dy is denoted $N(x, y)$, so $N(x, y) = 2xy + 1$. Since

$$\frac{\partial M}{\partial y} = 2y \quad \text{and} \quad \frac{\partial N}{\partial x} = 2y$$

the Test for Exactness says that the given differential equation is indeed exact (since $M_y = N_x$). This means that there exists a function $f(x, y)$ such that

$$\frac{\partial f}{\partial x} = M(x, y) = y^2 - 2x \quad \text{and} \quad \frac{\partial f}{\partial y} = N(x, y) = 2xy + 1$$

and once this function f is found, the general solution of the differential equation is simply

$$f(x, y) = c$$

(where c is an arbitrary constant). ■

Once a differential equation $M\,dx + N\,dy = 0$ is determined to be exact, the only task remaining is to find the function $f(x, y)$ such that $f_x = M$ and $f_y = N$. The method is simple: Integrate M with respect to x, integrate N with respect to y, and then "merge" the two resulting expressions to construct the desired function f.

Example 3: Solve the exact differential equation of Example 2:

$$(y^2 - 2x)\,dx + (2xy + 1)\,dy = 0$$

First, integrate $M(x, y) = y^2 - 2x$ with respect to x (and ignore the arbitrary "constant" of integration):

$$\int M(x, y)\,\partial x = \int (y^2 - 2x)\,\partial x = xy^2 - x^2$$

Next, integrate $N(x, y) = 2xy + 1$ with respect to y (and again ignore the arbitrary "constant" of integration):

$$\int N(x, y)\,\partial y = \int (2xy + 1)\,\partial y = xy^2 + y$$

Now, to "merge" these two expressions, write down each term exactly once, even if a particular term appears in both results. Here the two expressions contain the terms xy^2, $-x^2$, and y, so

$$f(x, y) = xy^2 - x^2 + y$$

(Note that the common term xy^2 is *not* written twice.) The general solution of the differential equation is $f(x, y) = c$, which in this case becomes

$$xy^2 - x^2 + y = c \quad \blacksquare$$

Example 4: Test the following equation for exactness and solve it if it is exact:

$$x(1 - \sin y)\,dy = (\cos x - \cos y - y)\,dx$$

First, bring the dx term over to the left-hand side to write the equation in standard form:

$$(y + \cos y - \cos x)\,dx + (x - x\sin y)\,dy = 0$$

Therefore, $M(x, y) = y + \cos y - \cos x$, and $N(x, y) = x - x \sin y$. Now, since

$$\frac{\partial M}{\partial y} = 1 - \sin y \quad \text{and} \quad \frac{\partial N}{\partial x} = 1 - \sin y$$

the Test for Exactness says that the differential equation is indeed exact (since $M_y = N_x$). To construct the function $f(x, y)$ such that $f_x = M$ and $f_y = N$, first integrate M with respect to x:

$$\int M(x, y)\,\partial x = \int (y + \cos y - \cos x)\,\partial x = xy + x\cos y - \sin x$$

Then integrate N with respect to y:

$$\int N(x, y)\,\partial y = \int (x - x\sin y)\,\partial y = xy + x\cos y$$

Writing all terms that appear in both these resulting expressions—without repeating any common terms—gives the desired function:

$$f(x, y) = xy + x\cos y - \sin x$$

The general solution of the given differential equation is therefore

$$xy + x\cos y - \sin x = c \quad \blacksquare$$

Example 5: Is the following equation exact?

$$(3xy - y^2)\,dx + x(x - y)\,dy = 0$$

Since

$$\frac{\partial M}{\partial y} = \frac{\partial}{\partial y}(3xy - y^2) = 3x - 2y$$

but

$$\frac{\partial N}{\partial x} = \frac{\partial}{\partial x}(x^2 - xy) = 2x - y$$

it is clear that $M_y \neq N_x$, so the Test for Exactness says that this equation is not exact. That is, there is no function $f(x, y)$ whose derivative with respect to x is $M(x, y) = 3xy - y^2$ and which at the same time has $N(x, y) = x(x - y)$ as its derivative with respect to y. $\blacksquare$

Example 6: Solve the IVP

$$(3x^2y - 1)\,dx + (x^3 + 6y - y^2)\,dy = 0$$

$$y(0) = 3$$

The differential equation is exact because

$$\frac{\partial M}{\partial y} = \frac{\partial}{\partial y}(3x^2y - 1) = 3x^2 \quad \text{and} \quad \frac{\partial N}{\partial x} = \frac{\partial}{\partial x}(x^3 + 6y - y^2) = 3x^2$$

Integrating M with respect to x gives

$$\int M(x, y)\,\partial x = \int (3x^2y - 1)\,\partial x = x^3y - x$$

and integrating N with respect to y yields

$$\int N(x,y)\,\partial y = \int (x^3 + 6y - y^2)\,\partial y = x^3y + 3y^2 - \tfrac{1}{3}y^3$$

Therefore, the function $f(x,y)$ whose total differential is the left-hand side of the given differential equation is

$$f(x,y) = x^3y - x + 3y^2 - \tfrac{1}{3}y^3$$

and the general solution is

$$x^3y - x + 3y^2 - \tfrac{1}{3}y^3 = c$$

The particular solution specified by the IVP must have $y = 3$ when $x = 0$; this condition determines the value of the constant c:

$$[x^3y - x + 3y^2 - \tfrac{1}{3}y^3]_{x=0,\,y=3} = c \Rightarrow 0 - 0 + 27 - 9 = c \Rightarrow 18 = c$$

Thus, the solution of the IVP is

$$x^3y - x + 3y^2 - \tfrac{1}{3}y^3 = 18 \qquad ■$$

Integrating Factors

If a differential equation of the form

$$M(x,y)\,dx + N(x,y)\,dy = 0 \qquad (*)$$

is not exact as written, then there exists a function $\mu(x,y)$ such that the equivalent equation obtained by multiplying both sides of (*) by μ,

$$(\mu M)\,dx + (\mu N)\,dy = 0$$

is exact. Such a function μ is called an **integrating factor** of the original equation and is guaranteed to exist if the given differential equation actually has a solution. *Integrating factors turn nonexact* *equations into exact ones.* The question is, how do you find an integrating factor? Two special cases will be considered.

Case 1: Consider the differential equation $M\,dx + N\,dy = 0$. If this equation is not exact, then M_y will not equal N_x; that is, $M_y - N_x \neq 0$. However, if

$$\frac{M_y - N_x}{N}$$

is a function of x only, let it be denoted by $\xi(x)$. Then

$$\mu(x) = e^{\int \xi(x)\,dx}$$

will be an integrating factor of the given differential equation.

Case 2: Consider the differential equation $M\,dx + N\,dy = 0$. If this equation is not exact, then M_y will not equal N_x; that is, $M_y - N_x \neq 0$. However, if

$$\frac{M_y - N_x}{-M}$$

is a function of y only, let it be denoted by $\psi(y)$. Then

$$\mu(y) = e^{\int \psi(y)\,dy}$$

will be an integrating factor of the given differential equation.

Example 7: The equation

$$(3xy - y^2)\,dx + x(x - y)\,dy = 0$$

is not exact, since

$$M_y = \frac{\partial}{\partial y}(3xy - y^2) = 3x - 2y \quad \text{but} \quad N_x = \frac{\partial}{\partial x}(x^2 - xy) = 2x - y$$

(recall Example 5). However, note that

$$\frac{M_y - N_x}{N} = \frac{(3x - 2y) - (2x - y)}{x(x - y)} = \frac{x - y}{x(x - y)} = \frac{1}{x}$$

is a function of x alone. Therefore, by Case 1,

$$e^{\int (1/x)\,dx} = e^{\ln x} = x$$

will be an integrating factor of the differential equation. Multiplying both sides of the given equation by $\mu = x$ yields

$$\underbrace{(3x^2y - xy^2)}_{\mu M = \overline{M}}\,dx + \underbrace{(x^3 - x^2y)}_{\mu N = \overline{N}}\,dy = 0$$

which *is* exact because

$$\frac{\partial \overline{M}}{\partial y} = 3x^2 - 2xy = \frac{\partial \overline{N}}{\partial x}$$

Solving this equivalent exact equation by the method described in the previous section, $\overline{M}$ is integrated with respect to x,

$$\int \overline{M}\,\partial x = \int (3x^2y - xy^2)\,\partial x = x^3y - \tfrac{1}{2}x^2y^2$$

and $\overline{N}$ is integrated with respect to y:

$$\int \overline{N}\,\partial y = \int (x^3 - x^2y)\,\partial y = x^3y - \tfrac{1}{2}x^2y^2$$

(with each "constant" of integration ignored, as usual). These calculations clearly give

$$x^3y - \tfrac{1}{2}x^2y^2 = c$$

as the general solution of the differential equation. ■

Example 8: The equation

$$(x + y)\sin y\,dx + (x \sin y + \cos y)\,dy = 0$$

is not exact, since

$$M_y = (x + y)\cos y + \sin y \quad \text{but} \quad N_x = \sin y$$

However, note that

$$\frac{M_y - N_x}{-M} = \frac{(x + y)\cos y + \sin y - \sin y}{-(x + y)\sin y} = -\frac{\cos y}{\sin y}$$

is a function of y alone (Case 2). Denote this function by $\psi(y)$; since

$$\int \psi(y)\,dy = -\int \frac{\cos y\,dy}{\sin y} = -\ln(\sin y)$$

the given differential equation will have

$$e^{\int \psi(y)\,dy} = e^{-\ln(\sin y)} = e^{\ln(\sin y)^{-1}} = (\sin y)^{-1}$$

as an integrating factor. Multiplying the differential equation through by $\mu = (\sin y)^{-1}$ yields

$$\underbrace{(x + y)\,dx}_{\mu M = \overline{M}} + \underbrace{\left(x + \frac{\cos y}{\sin y}\right)}_{\mu N = \overline{N}} dy = 0$$

which *is* exact because

$$\overline{M}_y = 1 = \overline{N}_x$$

To solve this exact equation, integrate $\overline{M}$ with respect to x and integrate $\overline{N}$ with respect to y, ignoring the "constant" of integration in each case:

$$\int \overline{M}\,\partial x = \int (x + y)\,\partial x = \tfrac{1}{2}x^2 + xy$$

$$\int \overline{N}\,\partial y = \int \left(x + \frac{\cos y}{\sin y}\right) \partial y = xy + \ln|\sin y|$$

These integrations imply that

$$\tfrac{1}{2}x^2 + xy + \ln|\sin y| = c$$

is the general solution of the differential equation. ■

Example 9: Solve the IVP

$$(3e^x y + x)\,dx + e^x\,dy = 0$$
$$y(0) = 1$$

The given differential equation is not exact, since

$$M_y = \frac{\partial}{\partial y}(3e^x y + x) = 3e^x \quad \text{but} \quad N_x = \frac{\partial}{\partial x}(e^x) = e^x$$

However, note that

$$\frac{M_y - N_x}{N} = \frac{3e^x - e^x}{e^x} = 2$$

which can be interpreted to be, say, a function of x only; that is, this last equation can be written as $\xi(x) \equiv 2$. Case 1 then says that

$$e^{\int \xi(x)\,dx} = e^{\int 2\,dx} = e^{2x}$$

will be an integrating factor. Multiplying both sides of the differential equation by $\mu(x) = e^{2x}$ yields

$$\underbrace{(3e^{3x}y + xe^{2x})}_{\mu M = \overline{M}}\,dx + \underbrace{(e^{3x})}_{\mu N = \overline{N}}\,dy = 0$$

which *is* exact because

$$\overline{M}_y = 3e^{3x} = \overline{N}_x$$

Now, since

$$\int \overline{M}\,\partial x = \int (3e^{3x}y + xe^{2x})\,\partial x = e^{3x}y + \tfrac{1}{2}xe^{2x} - \tfrac{1}{4}e^{2x}$$

and

$$\int \overline{N}\,\partial y = \int e^{3x}\,\partial y = e^{3x}y$$

(with the "constant" of integration suppressed in each calculation), the general solution of the differential equation is

$$e^{3x}y + \tfrac{1}{2}xe^{2x} - \tfrac{1}{4}e^{2x} = c$$

The value of the constant c is now determined by applying the initial condition $y(0) = 1$:

$$[e^{3x}y + \tfrac{1}{2}xe^{2x} - \tfrac{1}{4}e^{2x}]_{x=0,\,y=1} = c \Rightarrow \tfrac{3}{4} = c$$

Thus, the particular solution is

$$e^{3x}y + \tfrac{1}{2}xe^{2x} - \tfrac{1}{4}e^{2x} = \tfrac{3}{4}$$

which can be expressed explicitly as

$$y = \frac{3e^{-3x} + e^{-x}(1 - 2x)}{4} \quad ■$$

Example 10: Given that the nonexact differential equation

$$(5xy^2 - 2y)\,dx + (3x^2y - x)\,dy = 0$$

has an integrating factor of the form $\mu(x, y) = x^a y^b$ for some positive integers a and b, find the general solution of the equation.

Since there exist positive integers a and b such that $x^a y^b$ is an integrating factor, multiplying the differential equation through by this expression must yield an exact equation. That is,

$$\underbrace{(5x^{a+1}y^{b+2} - 2x^a y^{b+1})}_{\mu M = \overline{M}}\,dx + \underbrace{(3x^{a+2}y^{b+1} - x^{a+1}y^b)}_{\mu N = \overline{N}}\,dy = 0 \qquad (*)$$

is exact for some a and b. Exactness of this equation means

$$\overline{M}_y = \overline{N}_x$$

$$5(b+2)x^{a+1}y^{b+1} - 2(b+1)x^a y^b = 3(a+2)x^{a+1}y^{b+1} - (a+1)x^a y^b$$

By equating like terms in this last equation, it must be the case that

$$5(b+2) = 3(a+2) \quad \text{and} \quad 2(b+1) = a+1$$

The simultaneous solution of these equations is $a = 3$ and $b = 1$. Thus the integrating factor $x^a y^b$ is x^3y, and the exact equation $\overline{M}\,dx + \overline{N}\,dy = 0$ reads

$$(5x^4y^3 - 2x^3y^2)\,dx + (3x^5y^2 - x^4y)\,dy = 0$$

Now, since

$$\int \overline{M}\,\partial x = \int (5x^4y^3 - 2x^3y^2)\,\partial x = x^5y^3 - \tfrac{1}{2}x^4y^2$$

and

$$\int \overline{N}\,\partial y = \int (3x^5y^2 - x^4y)\,\partial y = x^5y^3 - \tfrac{1}{2}x^4y^2$$

(ignoring the "constant" of integration in each case), the general solution of the differential equation (*)—and hence the original differential equation—is clearly

$$x^5y^3 - \tfrac{1}{2}x^4y^2 = c \qquad \blacksquare$$

Separable Equations

Simply put, a differential equation is said to be **separable** if the variables can be separated. That is, a separable equation is one that can be written in the form

$$F(y)\,dy = G(x)\,dx$$

Once this is done, all that is needed to solve the equation is to integrate both sides. The method for solving separable equations can therefore be summarized as follows: *Separate the variables and integrate.*

Example 11: Solve the equation $2y\,dy = (x^2 + 1)\,dx$.

Since this equation is already expressed in "separated" form, just integrate:

$$2y\,dy = (x^2 + 1)\,dx$$

$$\int 2y\,dy = \int (x^2 + 1)\,dx$$

$$y^2 = \tfrac{1}{3}x^3 + x + c \qquad \blacksquare$$

Example 12: Solve the equation

$$x\,dx + \sec x \sin y\,dy = 0$$

This equation is separable, since the variables can be separated:

$$\sec x \sin y\,dy = -x\,dx$$

$$\sin y\,dy = -x \cos x\,dx$$

The integral of the left-hand side of this last equation is simply

$$\int \sin y\,dy = -\cos y$$

and the integral of the right-hand side is evaluated using integration by parts:

$$\begin{aligned}\int -x\cos x\,dx &= -\int x\cos x\,dx \\ &= -\left[x\sin x - \int \sin x\,dx\right] \\ &= -(x\sin x + \cos x + c)\end{aligned}$$

The solution of the differential equation is therefore

$$\cos y = x\sin x + \cos x + c \qquad \blacksquare$$

Example 13: Solve the IVP

$$\frac{dy}{dx} = \frac{x(e^{x^2}+2)}{6y^2}$$

$$y(0) = 1$$

The equation can be rewritten as follows:

$$6y^2\,dy = x(e^{x^2}+2)\,dx$$

Integrating both sides yields

$$\int 6y^2\,dy = \int x(e^{x^2}+2)\,dx$$

$$2y^3 = \tfrac{1}{2}e^{x^2} + x^2 + c$$

Since the initial condition states that $y = 1$ at $x = 0$, the parameter c can be evaluated:

$$[2y^3]_{y=1} = [\tfrac{1}{2}e^{x^2} + x^2 + c]_{x=0} \Rightarrow 2 = \tfrac{1}{2} + c \Rightarrow c = \tfrac{3}{2}$$

The solution of the IVP is therefore $2y^3 = \frac{1}{2}e^{x^2} + x^2 + \frac{3}{2}$ or

$$4y^3 = e^{x^2} + 2x^2 + 3 \qquad \blacksquare$$

Example 14: Find all solutions of the differential equation $(x^2 - 1)y^3\,dx + x^2\,dy = 0$.

Separating the variables and then integrating both sides gives

$$x^2\,dy = -(x^2 - 1)y^3\,dx$$

$$\frac{dy}{y^3} = \frac{1 - x^2}{x^2}\,dx \qquad (\dagger)$$

$$\int y^{-3}dy = \int (x^{-2} - 1)\,dx$$

$$-\tfrac{1}{2}y^{-2} = -x^{-1} - x - c$$

$$\frac{1}{2y^2} = \frac{1}{x} + x + c$$

Although the problem seems finished, there is another solution of the given differential equation that is not described by the family $\frac{1}{2}y^{-2} = x^{-1} + x + c$. In the separation step marked (†), both sides were divided by y^3. This operation prevented $y = 0$ from being derived as a solution (since division by zero is forbidden). It just so happens, however, that $y = 0$ *is* a solution of the given differential equation, as you can easily check (note: $y = 0 \Rightarrow dy = 0$).

Thus, the complete solution of this equation must include

both the family $\dfrac{1}{2y^2} = \dfrac{1}{x} + x + c$ *and*

the constant function $y = 0$

The lesson is clear:

If both sides of a separable differential equation are divided by some function $f(y)$ (that is, a function of the dependent variable) during the separation process, then a valid solution may be lost. As a final step, you must check whether the constant function $y = y_0$ [where $f(y_0) = 0$] is indeed a solution of the given differential equation. If it is, and if the family of solutions found by integrating both sides of the separated equation does not include this constant function, then this additional solution must be separately stated to complete the problem. ■

Example 15: Solve the equation

$$\frac{dy}{dx} = (1 + e^{-x})(y^2 - 1)$$

Separating the variables gives

$$\frac{dy}{y^2 - 1} = (1 + e^{-x})\,dx \qquad (\dagger)$$

(To achieve this separated form, note that both sides of the original equation were divided by $y^2 - 1$. Thus, the constant functions $y = 1$ and $y = -1$ may be lost as possible solutions; this will have to be checked later.) Integrating both sides of the separated equation yields

$$\int \frac{dy}{y^2 - 1} = \int (1 + e^{-x})\,dx$$

$$\int \left(\frac{-\frac{1}{2}}{y + 1} + \frac{\frac{1}{2}}{y - 1} \right) dy = \int (1 + e^{-x})\,dx$$

$$-\tfrac{1}{2} \ln |y + 1| + \tfrac{1}{2} \ln |y - 1| = x - e^{-x} + c$$

$$\ln \left| \frac{y - 1}{y + 1} \right| = 2(x - e^{-x} + c)$$

Now, both constant functions $y = 1$ and $y = -1$ *are* solutions of the original differential equation (as you can check by simply noting that $y = \pm 1 \Rightarrow dy/dx = 0$), and neither is described by the family above. Thus the complete set of solutions of the given differential equation includes

both the family $\ln\left|\frac{y-1}{y+1}\right| = 2(x - e^{-x} + c)$

and the constant functions $y = \pm 1$ ■

Example 16: Solve the differential equation $xy\,dx - (x^2 + 1)\,dy = 0$.

Separate the variables,

$$(x^2 + 1)\,dy = xy\,dx$$

$$\frac{dy}{y} = \frac{x\,dx}{x^2 + 1} \qquad (\dagger)$$

and integrate both sides:

$$\int \frac{dy}{y} = \int \frac{x\,dx}{x^2 + 1}$$

$$\ln|y| = \tfrac{1}{2}\ln(x^2 + 1) + c'$$

$$2\ln|y| = \ln(x^2 + 1) + \ln c$$

$$\ln(y^2) = \ln[c(x^2 + 1)]$$

$$y^2 = c(x^2 + 1)$$

Note that in the separation step (†), both sides were divided by y; thus, the solution $y = 0$ may have been lost. Direct substitution of the constant function $y = 0$ into the original differential equation shows that it is indeed a solution. However, the family $y^2 = c(x^2 + 1)$ already includes the function $y = 0$ (take $c = 0$), so it need not be separately mentioned. ■

Example 17: Find the curve $r = r(\theta)$ in polar coordinates that solves the IVP

$$\frac{dr}{d\theta} = -r \tan \theta$$

$$r(\pi) = 2$$

The given equation is separable, since it can be expressed in the separated form

$$\frac{dr}{r} = -\tan \theta \, d\theta \qquad (\dagger)$$

Now integrate both sides:

$$\int \frac{dr}{r} = \int \frac{-\sin \theta \, d\theta}{\cos \theta}$$

$$\ln |r| = \ln |\cos \theta| + c'$$

$$\ln |r| = \ln |\cos \theta| + \ln |c|$$

$$\ln |r| = \ln |c \cos \theta|$$

$$r = c \cos \theta$$

Since the solution curve is to pass through the point with polar coordinates $(r, \theta) = (2, \pi)$,

$$2 = c \cos \pi \Rightarrow c = -2$$

The solution of the IVP is therefore

$$r = -2 \cos \theta$$

This is a circle of diameter 2, tangent to the y axis at the origin; see Figure 5. Note: In the separation step (†), both sides were divided by r (which is the dependent variable here). However, even though $r = 0$ formally satisfies the differential equation, it clearly does not satisfy the initial condition $r(\pi) = 2$. ■

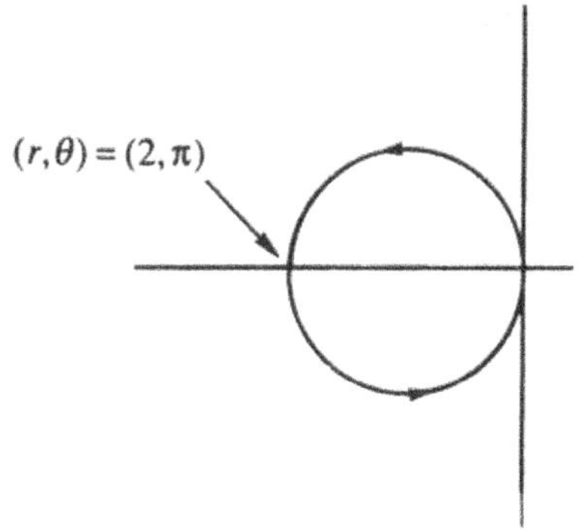

■ Figure 5 ■

Homogeneous Equations

A function $f(x, y)$ is said to be **homogeneous of degree *n*** if the equation

$$f(zx, zy) = z^n f(x, y)$$

holds for all x, y, and z (for which both sides are defined).

Example 18: The function $f(x, y) = x^2 + y^2$ is homogeneous of degree 2, since

$$f(zx, zy) = (zx)^2 + (zy)^2 = z^2(x^2 + y^2) = z^2 f(x, y) \quad ■$$

Example 19: The function $f(x, y) = \sqrt{x^8 - 3x^2y^6}$ is homogeneous of degree 4, since

$$\begin{aligned} f(zx, zy) &= \sqrt{(zx)^8 - 3(zx)^2(zy)^6} = \sqrt{z^8(x^8 - 3x^2y^6)} \\ &= \sqrt{z^8}\,\sqrt{x^8 - 3x^2y^6} = z^4 f(x, y) \quad ■ \end{aligned}$$

Example 20: The function $f(x, y) = 2x + y$ is homogeneous of degree 1, since

$$f(zx, zy) = 2(zx) + (zy) = z(2x + y) = z^1 f(x, y) \quad \blacksquare$$

Example 21: The function $f(x, y) = x^3 - y^2$ is not homogeneous, since

$$f(zx, zy) = (zx)^3 - (zy)^2 = z^3x^3 - z^2y^2$$

which does not equal $z^n f(x, y)$ for any n. ■

Example 22: The function $f(x, y) = x^3 \sin(y/x)$ is homogeneous of degree 3, since

$$f(zx, zy) = (zx)^3 \sin\frac{zy}{zx} = z^3\left(x^3 \sin\frac{y}{x}\right) = z^3 f(x, y) \quad \blacksquare$$

A first-order differential equation

$$M(x, y)\,dx + N(x, y)\,dy = 0$$

is said to be **homogeneous** if $M(x, y)$ and $N(x, y)$ are both homogeneous functions of the same degree.

Example 23: The differential equation

$$(x^2 - y^2)\,dx + xy\,dy = 0$$

is homogeneous because both $M(x, y) = x^2 - y^2$ and $N(x, y) = xy$ are homogeneous functions of the same degree (namely, 2). ■

The method for solving homogeneous equations follows from this fact:

> The substitution $y = xv$ (and therefore $dy = x\,dv + v\,dx$) transforms a homogeneous equation into a separable one.

Example 24: Solve the equation $(x^2 - y^2)\,dx + xy\,dy = 0$.

This equation is homogeneous, as observed in Example 23. Thus to solve it, make the substitutions $y = xv$ and $dy = x\,dv + v\,dx$:

$$[x^2 - (xv)^2]\,dx + [x(xv)](x\,dv + v\,dx) = 0$$

$$(x^2 - x^2v^2)\,dx + x^3v\,dv + x^2v^2\,dx = 0$$

$$x^2dx + x^3v\,dv = 0$$

$$dx + xv\,dv = 0$$

This final equation is now separable (which was the intention). Proceeding with the solution,

$$v\,dv = -\frac{dx}{x}$$

$$\int v\,dv = \int -\frac{dx}{x}$$

$$\tfrac{1}{2}v^2 = -\ln|x| + c'$$

Therefore, the solution of the separable equation involving x and v can be written

$$\tfrac{1}{2}v^2 = \ln\left|\frac{c}{x}\right|$$

To give the solution of the original differential equation (which involved the variables x and y), simply note that

$$y = xv \Rightarrow v = \frac{y}{x}$$

Replacing v by y/x in the solution above gives the final result:

$$\frac{1}{2}\left(\frac{y}{x}\right)^2 = \ln\left|\frac{c}{x}\right| \Rightarrow y^2 = 2x^2 \ln\left|\frac{c}{x}\right|$$

This is the general solution of the original differential equation. ■

Example 25: Solve the IVP

$$\begin{aligned} 2(x + 2y)\,dx + (y - x)\,dy &= 0 \\ y(1) &= 0 \end{aligned}$$

Since the functions

$$M(x, y) = 2(x + 2y) \quad \text{and} \quad N(x, y) = y - x$$

are both homogeneous of degree 1, the differential equation is homogeneous. The substitutions $y = xv$ and $dy = x\,dv + v\,dx$ transform the equation into

$$2(x + 2xv)\,dx + (xv - x)(x\,dv + v\,dx) = 0$$

which simplifies as follows:

$$\begin{aligned} 2x\,dx + 4xv\,dx + x^2v\,dv - x^2\,dv + xv^2\,dx - xv\,dx &= 0 \\ (2x + 3xv + xv^2)\,dx + (x^2v - x^2)\,dv &= 0 \\ x(2 + 3v + v^2)\,dx + x^2(v - 1)\,dv &= 0 \\ (2 + 3v + v^2)\,dx + x(v - 1)\,dv &= 0 \end{aligned}$$

The equation is now separable. Separating the variables and integrating gives

$$x(v - 1)\,dv = -(2 + 3v + v^2)\,dx$$

$$\frac{v - 1}{v^2 + 3v + 2}\,dv = -\frac{dx}{x} \qquad (\dagger)$$

The integral of the left-hand side is evaluated after performing a partial fraction decomposition:

$$\frac{v - 1}{v^2 + 3v + 2} = \frac{v - 1}{(v + 1)(v + 2)} = \frac{-2}{v + 1} + \frac{3}{v + 2}$$

Therefore,

$$\begin{aligned}\int \frac{v - 1}{v^2 + 3v + 2}\,dv &= \int \left(\frac{-2}{v + 1} + \frac{3}{v + 2}\right) dv \\ &= -2 \ln|v + 1| + 3 \ln|v + 2| \\ &= \ln|(v + 1)^{-2}(v + 2)^3|\end{aligned}$$

The right-hand side of (†) immediately integrates to

$$\int -\frac{dx}{x} = -\ln|x| + c' = \ln|cx^{-1}|$$

Therefore, the solution to the separable differential equation (†) is

$$(v + 1)^{-2}\,(v + 2)^3 = cx^{-1}$$

Now, replacing v by y/x gives

$$\left(\frac{y}{x} + 1\right)^{-2}\left(\frac{y}{x} + 2\right)^3 = cx^{-1}$$

as the general solution of the given differential equation. Applying the initial condition $y(1) = 0$ determines the value of the constant c:

$$\left(\frac{0}{1} + 1\right)^{-2}\left(\frac{0}{1} + 2\right)^{3} = c \cdot 1^{-1} \Rightarrow 8 = c$$

Thus, the particular solution of the IVP is

$$\left(\frac{y}{x} + 1\right)^{-2}\left(\frac{y}{x} + 2\right)^{3} = 8x^{-1}$$

which can be simplified to

$$(2x + y)^3 = 8(x + y)^2$$

as you can check.

Technical note: In the separation step (†), both sides were divided by $(v + 1)(v + 2)$, and $v = -1$ and $v = -2$ were lost as solutions. These need not be considered, however, because even though the equivalent functions $y = -x$ and $y = -2x$ do indeed satisfy the given differential equation, they are inconsistent with the initial condition. ■

Linear Equations

A first-order differential equation is said to be **linear** if it can be expressed in the form

$$y' + P(x)y = Q(x)$$

where P and Q are functions of x. The method for solving such equations is similar to the one used to solve nonexact equations. There, the nonexact equation was multiplied by an integrating factor, which then made it easy to solve (because the equation became exact).

To solve a first-order linear equation, first rewrite it (if necessary) in the standard form above; then multiply both sides by the **integrating factor**

$$\mu(x) = e^{\int P\,dx}$$

The resulting equation,

$$\mu \frac{dy}{dx} + \mu Py = \mu Q \qquad (*)$$

is then easy to solve, not because it's exact, but because the left-hand side collapses:

$$\begin{aligned} \mu \frac{dy}{dx} + \mu Py &= \mu \frac{dy}{dx} + e^{\int P\,dx} \cdot Py \\ &= \mu \frac{dy}{dx} + y \frac{d}{dx}(e^{\int P\,dx}) \\ &= \mu \frac{dy}{dx} + y \frac{d\mu}{dx} \\ &= \frac{d}{dx}(\mu y) \end{aligned}$$

Therefore, equation (*) becomes

$$\frac{d}{dx}(\mu y) = \mu Q$$

making it susceptible to an integration, which gives the solution:

$$\mu y = \int (\mu Q)\,dx$$

Do not memorize this equation for the solution; memorize the steps needed to get there.

Example 26: Solve the differential equation

$$y' + 2xy = x$$

The equation is already expressed in standard form, with $P(x) = 2x$ and $Q(x) = x$. Multiplying both sides by

$$\mu(x) = e^{\int P\,dx} = e^{\int 2x\,dx} = e^{x^2}$$

transforms the given differential equation into

$$e^{x^2}y' + 2xe^{x^2}y = xe^{x^2}$$

$$\frac{d}{dx}(e^{x^2}y) = xe^{x^2}$$

Notice how the left-hand side collapses into $(\mu y)'$; as shown above, *this will always happen.* Integrating both sides gives the solution:

$$e^{x^2}y = \int xe^{x^2}\,dx$$

$$e^{x^2}y = \tfrac{1}{2}e^{x^2} + c$$

$$y = \tfrac{1}{2} + ce^{-x^2} \quad \blacksquare$$

Example 27: Solve the IVP

$$y' + \frac{1}{x}y = \sin x$$

$$y(\pi) = 1$$

Note that the differential equation is already in standard form. Since $P(x) = 1/x$, the integrating factor is

$$\mu(x) = e^{\int P\,dx} = e^{\int (1/x)\,dx} = e^{\ln x} = x$$

Multiplying both sides of the standard-form differential equation by $\mu = x$ gives

$$xy' + y = x \sin x$$

$$(xy)' = x \sin x$$

Again, note how the left-hand side automatically collapses into $(\mu y)'$. Integrating both sides yields the general solution:

$$xy = \int x \sin x \, dx$$

$$xy = -x \cos x + \sin x + c$$

Applying the initial condition $y(\pi) = 1$ determines the constant c:

$$\pi \cdot 1 = -\pi \cos \pi + \sin \pi + c \Rightarrow c = 0$$

Thus the desired particular solution is

$$xy = -x \cos x + \sin x$$

or, since x cannot equal zero (note the coefficient $P(x) = 1/x$ in the given differential equation),

$$y = \frac{\sin x}{x} - \cos x \qquad \blacksquare$$

Example 28: Solve the linear differential equation

$$x^2 \frac{dy}{dx} = -2y$$

First, rewrite the equation in standard form:

$$\frac{dy}{dx} + \frac{2}{x^2} y = 0 \qquad (*)$$

Since the integrating factor here is

$$\mu(x) = e^{\int P\,dx} = e^{\int (2/x^2)\,dx} = e^{-2/x}$$

multiply both sides of the standard-form equation (*) by $\mu = e^{-2/x}$,

$$e^{-2/x}\frac{dy}{dx} + \frac{2}{x^2}e^{-2/x}y = 0$$

collapse the left-hand side,

$$\frac{d}{dx}(e^{-2/x}y) = 0$$

and integrate:

$$e^{-2/x}y = c$$

Thus the general solution of the differential equation can be expressed explicitly as

$$y = ce^{2/x} \quad ■$$

Example 29: Find the general solution of each of the following equations:

(a) $\dfrac{dy}{dx} - \dfrac{4}{x}y = 0$

(b) $\dfrac{dy}{dx} - \dfrac{4}{x}y = x^4$

Both equations are linear equations in standard form, with $P(x) = -4/x$. Since

$$\int P\,dx = \int -\frac{4}{x}\,dx = -4\ln|x| = \ln(x^{-4})$$

the integrating factor will be

$$\mu(x) = e^{\int P\,dx} = e^{\ln(x^{-4})} = x^{-4}$$

for both equations. Multiplying through by $\mu = x^{-4}$ yields

$$\frac{d}{dx}(x^{-4}y) = 0 \quad \text{for equation (a)}$$

$$\frac{d}{dx}(x^{-4}y) = 1 \quad \text{for equation (b)}$$

Integrating each of these resulting equations gives the general solutions:

$$x^{-4}y = c \Rightarrow y = cx^4 \qquad \text{for equation (a)}$$
$$x^{-4}y = x + c \Rightarrow y = cx^4 + x^5 \quad \text{for equation (b)} \qquad \blacksquare$$

Example 30: Sketch the integral curve of

$$(1 + x^2)y' = x(1 - y)$$

which passes through the origin.

The first step is to rewrite the differential equation in standard form:

$$(1 + x^2)y' + xy = x$$

$$y' + \frac{x}{1 + x^2}y = \frac{x}{1 + x^2} \qquad (*)$$

Since

$$\int P\,dx = \int \frac{x}{1 + x^2}\,dx = \tfrac{1}{2}\ln(1 + x^2)$$

the integrating factor is

$$\mu(x) = e^{\int P\,dx} = e^{(1/2)\ln(1+x^2)} = e^{\ln(1+x^2)^{1/2}} = (1 + x^2)^{1/2}$$

Multiplying both sides of the standard-form equation (*) by $\mu = (1 + x^2)^{1/2}$ gives

$$(1 + x^2)^{1/2}y' + \frac{x}{(1 + x^2)^{1/2}}y = \frac{x}{(1 + x^2)^{1/2}}$$

As usual, the left-hand side collapses into $(\mu y)'$,

$$\frac{d}{dx}[(1 + x^2)^{1/2}y] = \frac{x}{(1 + x^2)^{1/2}}$$

and an integration gives the general solution:

$$(1 + x^2)^{1/2}y = \int x(1 + x^2)^{-1/2}\,dx$$

$$(1 + x^2)^{1/2}y = (1 + x^2)^{1/2} + c$$

$$y = 1 + c(1 + x^2)^{-1/2}$$

To find the particular curve of this family that passes through the origin, substitute $(x, y) = (0, 0)$ and evaluate the constant c:

$$0 = 1 + c(1 + 0)^{-1/2} \Rightarrow c = -1$$

Therefore, the desired integral curve is

$$y = 1 - \frac{1}{\sqrt{1 + x^2}}$$

which is sketched in Figure 6. ■

Example 31: An object moves along the x axis in such a way that its position at time $t > 0$ is governed by the linear differential equation

$$\frac{dx}{dt} + (t - t^{-1})x = t^2$$

If the object was at position $x = 2$ at time $t = 1$, where will it be at time $t = 3$?

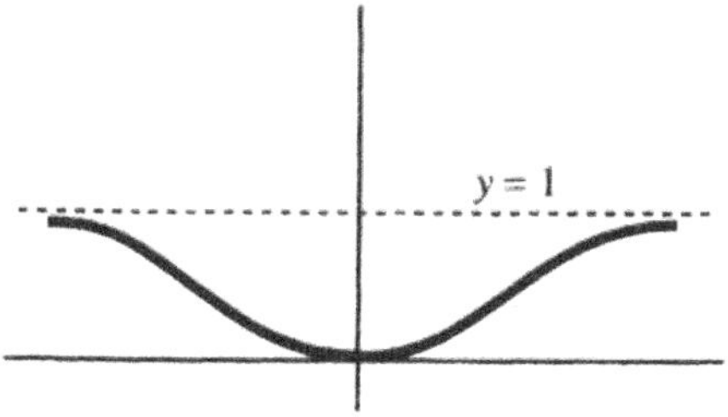

■ Figure 6 ■

Rather than having x as the independent variable and y as the dependent one, in this problem t is the independent variable and x is the dependent one. Thus, the solution will not be of the form "y = some function of x" but will instead be "x = some function of t."

The equation is in the standard form for a first-order linear equation, with $P = t - t^{-1}$ and $Q = t^2$. Since

$$\int P\,dt = \int (t - t^{-1})\,dt = \tfrac{1}{2}t^2 - \ln t$$

the integrating factor is

$$\mu(t) = e^{\int P\,dt} = e^{(t^2/2) - \ln t} = e^{t^2/2}e^{-\ln t} = e^{t^2/2}e^{\ln t^{-1}} = t^{-1}e^{t^2/2}$$

Multiplying both sides of the differential equation by this integrating factor transforms it into

$$t^{-1}e^{t^2/2}\frac{dx}{dt} + (1 - t^{-2})e^{t^2/2}x = te^{t^2/2}$$

As usual, the left-hand side automatically collapses,

$$\frac{d}{dt}[t^{-1}e^{t^2/2}x] = te^{t^2/2}$$

and an integration yields the general solution:

$$t^{-1}e^{t^2/2}x = \int te^{t^2/2}\,dt$$

$$= e^{t^2/2} + c$$

$$x = t(1 + ce^{-t^2/2})$$

Now, since the condition "$x = 2$ at $t = 1$" is given, this is actually an IVP, and the constant c can be evaluated:

$$2 = 1 \cdot (1 + ce^{-1/2}) \Rightarrow ce^{-1/2} = 1 \Rightarrow c = e^{1/2}$$

Thus, the position x of the object as a function of time t is given by the equation

$$x = t[1 + e^{(1-t^2)/2}]$$

and therefore, the position at time $t = 3$ is

$$x|_{t=3} = 3(1 + e^{-4})$$

which is approximately 3.055. ■

Bernoulli's Equation

The differential equation

$$y' + P(x)y = Q(x)y^n$$

is known as **Bernoulli's equation.** If $n = 0$, Bernoulli's equation reduces immediately to the standard form first-order linear equation:

$$y' + P(x)y = Q(x)$$

If $n = 1$, the equation can also be written as a linear equation:

$$y' + P(x)y = Q(x)y \Rightarrow y' + [P(x) - Q(x)]y = 0$$

However, if n is not 0 or 1, then Bernoulli's equation is not linear. Nevertheless, it can be *transformed* into a linear equation (and therefore solved by the method of the preceding section) by first multiplying through by y^{-n},

$$y^{-n}y' + P(x)y^{1-n} = Q(x)$$

and then introducing the substitutions

$$\begin{cases} w = y^{1-n} \\ w' = (1-n)y^{-n}y' \end{cases}$$

The equation above then becomes

$$\frac{1}{1-n}w' + P(x)w = Q(x)$$

which is linear in w (since $n \neq 1$).

Example 32: Solve the equation

$$y' + xy = xy^3$$

Note that this fits the form of the Bernoulli equation with $n = 3$. Therefore, the first step in solving it is to multiply through by $y^{-n} = y^{-3}$:

$$y^{-3}y' + xy^{-2} = x \qquad (*)$$

Now for the substitutions; the equations

$$\begin{cases} w = y^{1-3} = y^{-2} \\ w' = -2y^{-3}y' \end{cases}$$

transform (*) into

$$-\tfrac{1}{2}w' + xw = x$$

or, in standard form,

$$w' - 2xw = -2x \qquad (**)$$

Notice that the substitutions were successful in transforming the Bernoulli equation into a linear equation (just as they were designed to be). To solve the resulting linear equation, first determine the integrating factor:

$$\mu(x) = e^{\int P\,dx} = e^{\int -2x\,dx} = e^{-x^2}$$

Multiplying (**) through the $\mu = e^{-x^2}$ yields

$$e^{-x^2}w' - 2xe^{-x^2}w = -2xe^{-x^2}$$

$$(e^{-x^2}w)' = -2xe^{-x^2}$$

And an integration gives

$$e^{-x^2}w = \int -2xe^{-x^2}\,dx$$

$$e^{-x^2}w = e^{-x^2} + c$$

$$w = 1 + ce^{x^2}$$

The final step is simply to undo the substitution $w = y^{-2}$. The solution to the original differential equation is therefore

$$\frac{1}{y^2} = 1 + ce^{x^2} \qquad \blacksquare$$

Second-order differential equations involve the second derivative of the unknown function (and, quite possibly, the first derivative as well) but no derivatives of higher order. For nearly every second-order equation encountered in practice, the general solution will contain two arbitrary constants, so a second-order IVP must include two initial conditions.

Linear Combinations and Linear Independence

Given two functions $y_1(x)$ and $y_2(x)$, any expression of the form

$$c_1y_1 + c_2y_2$$

where c_1 and c_2 are constants, is called a **linear combination** of y_1 and y_2. For example, if $y_1 = e^x$ and $y_2 = x^2$, then

$$5e^x + 7x^2, \quad e^x - 2x^2, \quad \text{and} \quad -e^x + x^2$$

are all particular linear combinations of y_1 and y_2. So the idea of a linear combination of two functions is this: Multiply the functions by whatever constants you wish; then add the products.

Example 1: Is $y = 2x$ a linear combination of the functions $y_1 = x$ and $y_2 = x^2$?

Any expression that can be written in the form

$$c_1x + c_2x^2$$

is a linear combination of x and x^2. Since $y = 2x$ fits this form by taking $c_1 = 2$ and $c_2 = 0$, $y = 2x$ is indeed a linear combination of x and x^2. ■

Example 2: Consider the three functions $y_1 = \sin x$, $y_2 = \cos x$, and $y_3 = \sin(x + 1)$. Show that y_3 is a linear combination of y_1 and y_2.

The addition formula for the sine function says

$$\sin(x + 1) = \cos 1 \sin x + \sin 1 \cos x$$

Note that this fits the form of a linear combination of $\sin x$ and $\cos x$,

$$c_1 \sin x + c_2 \cos x$$

by taking $c_1 = \cos 1$ and $c_2 = \sin 1$. ■

Example 3: Can the function $y = x^3$ be written as a linear combination of the functions $y_1 = x$ and $y_2 = x^2$?

If the answer were yes, then there would be constants c_1 and c_2 such that the equation

$$x^3 = c_1x + c_2x^2 \quad (*)$$

holds true for *all* values of x. Letting $x = 1$ in this equation gives

$$1 = c_1 + c_2$$

and letting $x = -1$ gives

$$-1 = -c_1 + c_2$$

Adding these last two equations gives $0 = 2c_2$, so $c_2 = 0$. And since $c_2 = 0$, c_1 must equal 1. Thus, the general linear combination (*) reduces to

$$x^3 = x$$

which clearly does *not* hold for all values of x. Therefore, it is not possible to write $y = x^3$ as a linear combination of $y_1 = x$ and $y_2 = x^2$. ■

One more definition: Two functions y_1 and y_2 are said to be **linearly independent** if neither function is a constant multiple of the other. For example, the functions $y_1 = x^3$ and $y_2 = 5x^3$ are *not* linearly independent (they're **linearly dependent**), since y_2 is clearly a constant multiple of y_1. Checking that two functions are dependent is easy; checking that they're independent takes a little more work.

Example 4: Are the functions $y_1(x) = \sin x$ and $y_2(x) = \cos x$ linearly independent?

If they weren't, then y_1 would be a constant multiple of y_2; that is, the equation

$$\sin x = c \cos x$$

would hold for some constant c and for all x. But substituting $x = \pi/2$, for example, yields the absurd statement $1 = 0$. Therefore, the above equation cannot be true: $y_1 = \sin x$ is *not* a constant multiple of $y_2 = \cos x$; thus, these functions are indeed linearly independent. ■

Example 5: Are the functions $y_1 = e^x$ and $y_2 = x$ linearly independent?

If they weren't, then y_1 would be a constant multiple of y_2; that is, the equation

$$e^x = cx$$

would hold for some constant c and for all x. But this cannot happen, since substituting $x = 0$, for example, yields the absurd statement $1 = 0$. Therefore, $y_1 = e^x$ is *not* a constant multiple of $y_2 = x$; these two functions are linearly independent. ■

Example 6: Are the functions $y_1 = xe^x$ and $y_2 = e^x$ linearly independent?

A hasty conclusion might be to say no because y_1 is a multiple of y_2. But y_1 is not a *constant* multiple of y_2, so these functions truly are independent. (You may find it instructive to prove they're independent by the same kind of argument used in the previous two examples.) ■

Linear Equations

Recall that the order of a differential equation is the order of the highest derivative appearing in the equation. Thus, a second-order differential equation is one that involves the second derivative of the unknown function but no higher derivatives.

A second-order **linear** differential equation is one that can be written in the form

$$a(x)y'' + b(x)y' + c(x)y = d(x)$$

where $a(x)$ is not identically zero. [For if $a(x)$ were identically zero, then the equation really wouldn't contain a second-derivative term, so it wouldn't be a second-order equation.] If $a(x) \neq 0$, then both sides of the equation can be divided through by $a(x)$ and the resulting equation written in the form

$$y'' + p(x)y' + q(x)y = r(x)$$

It is a fact that as long as the functions p, q, and r are continuous on some interval, then the equation will indeed have a solution (on that interval), which will in general contain *two* arbitrary constants (as you should expect for the general solution of a *second*-order differential equation). What will this solution look like? There is no explicit formula that will give the solution in all cases, only various methods that work depending on the properties of the coefficient functions p, q, and r. But there is something definitive—and very important—that *can* be said about second-order linear equations.

Homogeneous Equations

There are two definitions of the term "homogeneous differential equation." One definition (already encountered) calls a first-order equation of the form

$$M(x, y)\,dx + N(x, y)\,dy = 0$$

homogeneous if M and N are both homogeneous functions of the same degree. The second definition—and the one which you'll see much more often—states that a differential equation (of *any* order) is **homogeneous** if once all the terms involving the unknown function are collected together on one side of the equation, the other side is identically zero. For example,

$$y'' - 2y' + y = 0 \quad \text{is homogeneous}$$

but

$$y'' - 2y' + y = x \quad \text{is not}$$

For the rest of this book, the term "homogeneous" will refer only to this latter definition.

The nonhomogeneous equation

$$a(x)y'' + b(x)y' + c(x)y = d(x) \qquad (*)$$

can be turned into a homogeneous one simply by replacing the right-hand side by 0:

$$a(x)y'' + b(x)y' + c(x)y = 0 \qquad (**)$$

Equation (**) is called the **homogeneous equation corresponding to the nonhomogeneous equation** (*). There is an important connection between the solution of a nonhomogeneous linear equation and the solution of its corresponding homogeneous equation. The two principal results of this section are given below:

Theorem A. If $y_1(x)$ and $y_2(x)$ are linearly independent solutions of the linear homogeneous equation (**), then *every* solution is a linear combination of y_1 and y_2. That is, the general solution of the linear homogeneous equation is

$$y = c_1y_1 + c_2y_2$$

Theorem B. If $\bar{y}(x)$ is any particular solution of the linear nonhomogeneous equation (*), and if $y_h(x)$ is the general solution of the corresponding homogeneous equation, then the general solution of the linear nonhomogeneous equation is

$$y = y_h + \bar{y}$$

That is,

general solution of linear nonhomogeneous equation = general solution of corresponding homogeneous equation + particular solution of given nonhomogeneous equation

[Note: The general solution of the corresponding homogeneous equation, which has been denoted here by y_h, is sometimes called the **complementary function** of the nonhomogeneous equation (*).] Theorem A can be generalized to homogeneous linear equations of any order, while Theorem B as written holds true for linear equations of any order. You can see Theorem B illustrated for a first-order equation by looking back at Example 29 on page 56. Theorems A and B are perhaps the most important theoretical facts about linear differential equations—definitely worth memorizing.

Example 7: The differential equation

$$y'' - 2y' + y = 0$$

is satisfied by the functions

$$y_1 = e^x \quad \text{and} \quad y_2 = xe^x$$

Verify that any linear combination of y_1 and y_2 is also a solution of this equation. What is its general solution?

Every linear combination of $y_1 = e^x$ and $y_2 = xe^x$ looks like this:

$$y = c_1e^x + c_2xe^x$$

for some constants c_1 and c_2. To verify that this satisfies the differential equation, just substitute. If $y = c_1e^x + c_2xe^x$, then

$$y' = c_1e^x + c_2e^x(x + 1)$$

$$y'' = c_1e^x + c_2e^x(x + 2)$$

Substituting these expressions into the left-hand side of the given differential equation gives

$$\begin{aligned} y'' - 2y' + y &= [c_1e^x + c_2e^x(x + 2)] - 2[c_1e^x + c_2e^x(x + 1)] \\ &\qquad + [c_1e^x + c_2xe^x] \\ &= c_1(e^x - 2e^x + e^x) + c_2[e^x(x + 2) - 2e^x(x + 1) + xe^x] \\ &= 0 \quad \checkmark \end{aligned}$$

Thus, any linear combination of $y_1 = e^x$ and $y_2 = xe^x$ does indeed satisfy the differential equation. Now, since $y_1 = e^x$ and $y_2 = xe^x$ are linearly independent (see Example 6), Theorem A says that the general solution of the equation is

$$y = c_1e^x + c_2xe^x \quad \blacksquare$$

Example 8: Verify that $y = 4x - 5$ satisfies the equation

$$y'' + 5y' + 4y = 16x$$

Then, given that $y_1 = e^{-x}$ and $y_2 = e^{-4x}$ are solutions of the corresponding homogeneous equation, write the general solution of the given nonhomogeneous equation.

First, to verify that $y = 4x - 5$ is a particular solution of the nonhomogeneous equation, just substitute. If $y = 4x - 5$, then $y' = 4$ and $y'' = 0$, so the left-hand side of the equation becomes

$$\begin{aligned} y'' + 5y' + 4y &= 0 + 5(4) + 4(4x - 5) \\ &= 16x \quad \checkmark \end{aligned}$$

Now, since the functions $y_1 = e^{-x}$ and $y_2 = e^{-4x}$ are linearly independent (because neither is a constant multiple of the other), Theorem A says that the general solution of the corresponding homogeneous equation is

$$y_h = c_1e^{-x} + c_2e^{-4x}$$

Theorem B then says

$$y = \underbrace{c_1e^{-x} + c_2e^{-4x}}_{y_h} + \underbrace{4x - 5}_{\bar{y}}$$

is the general solution of the given nonhomogeneous equation. ■

Example 9: Verify that both $y_1 = \sin x$ and $y_2 = \cos x$ satisfy the homogeneous differential equation $y'' + y = 0$. What then is the general solution of the nonhomogeneous equation $y'' + y = x$?

If $y_1 = \sin x$, then $y_1'' = -\sin x$, so $y_1'' + y_1$ does indeed equal zero. Similarly, if $y_2 = \cos x$, then $y_2'' = -\cos x$, so $y_2'' + y_2$ is also zero, as desired. Since $y_1 = \sin x$ and $y_2 = \cos x$ are linearly independent (see

Example 4), Theorem A says that the general solution of the homogeneous equation $y'' + y = 0$ is

$$y_h = c_1 \sin x + c_2 \cos x$$

Now, to solve the given nonhomogeneous equation, all that is needed is any particular solution. By inspection, you can see that $\bar{y} = x$ satisfies $y'' + y = x$. Therefore, according to Theorem B, the general solution of this nonhomogeneous equation is

$$y = c_1 \sin x + c_2 \cos x + x \qquad \blacksquare$$

Homogeneous Linear Equations with Constant Coefficients

The general second-order homogeneous linear differential equation has the form

$$a(x)y'' + b(x)y' + c(x)y = 0 \qquad [a(x) \not\equiv 0]$$

If $a(x)$, $b(x)$, and $c(x)$ are actually constants, $a(x) \equiv a \neq 0$, $b(x) \equiv b$, $c(x) \equiv c$, then the equation becomes simply

$$ay'' + by' + cy = 0$$

This is the general second-order homogeneous linear equation **with constant coefficients.**

Theorem A above says that the general solution of this equation is the general linear combination of any two linearly independent solutions. So how are these two linearly independent solutions found? The following example will illustrate the fundamental idea.

Example 10: Solve the differential equation $y'' - y' - 2y = 0$.

The trick is to substitute $y = e^{mx}$ (m a constant) into the equation; you will see shortly why this approach works. If $y = e^{mx}$, then $y' = me^{mx}$ and $y'' = m^2e^{mx}$, so the differential equation becomes

$$m^2e^{mx} - me^{mx} - 2e^{mx} = 0$$

The term e^{mx} can be factored out and immediately canceled (since e^{mx} never equals zero):

$$e^{mx}(m^2 - m - 2) = 0$$

$$m^2 - m - 2 = 0$$

This quadratic polynomial equation can be solved by factoring:

$$m^2 - m - 2 = 0$$

$$(m + 1)(m - 2) = 0$$

$$m = -1, 2$$

Now, recall that the solution began by writing $y = e^{mx}$. Since the values of m have now been found ($m = -1, m = 2$), both

$$y = e^{-x} \quad \text{and} \quad y = e^{2x}$$

are solutions. Since these functions are linearly independent (neither is a constant multiple of the other), Theorem A says that the general linear combination

$$y = c_1e^{-x} + c_2e^{2x}$$

is the general solution of the differential equation. ■

Note carefully that the solution of the homogeneous differential equation

$$ay'' + by' + cy = 0$$

depends entirely on the roots of the **auxiliary polynomial** equation that results from substituting $y = e^{mx}$ and then canceling out the e^{mx} term. Once the roots of this auxiliary polynomial equation are found, you can immediately write down the general solution of the given differential equation. Also note that a *second*-order linear homogeneous differential equation with constant coefficients will always give rise to a *second*-degree auxiliary polynomial equation, that is, to a *quadratic* polynomial equation.

The roots of any quadratic equation

$$am^2 + bm + c = 0 \qquad (a \neq 0)$$

are given by the famous **quadratic formula**

$$m = \frac{-b \pm \sqrt{b^2 - 4ac}}{2a}$$

The quantity under the square root sign, $b^2 - 4ac$, is called the **discriminant** of the equation, and its sign determines the nature of the roots. There are exactly three cases to consider.

Case 1: The discriminant is positive.

In this case, the roots are real and distinct. If the two roots are denoted m_1 and m_2, then the general solution of the differential equation is

$$y = c_1 e^{m_1 x} + c_2 e^{m_2 x}$$

Case 2: The discriminant is zero.

In this case, the roots are real and identical; that is, the polynomial equation has a double (repeated) root. If this double root is denoted simply by m, then the general solution of the differential equation is

$$y = c_1 e^{mx} + c_2 x e^{mx}$$

Case 3: The discriminant is negative.

In this case, the roots are distinct conjugate complex numbers, $r \pm si$. The general solution of the differential equation is then

$$y = e^{rx}\,(c_1 \cos sx + c_2 \sin sx)$$

So here's the process: Given a second-order homogeneous linear differential equation with constant coefficients ($a \neq 0$),

$$ay'' + by' + cy = 0$$

immediately write down the corresponding auxiliary quadratic polynomial equation

$$am^2 + bm + c = 0$$

(found by simply replacing y'' by m^2, y' by m, and y by 1). Determine the roots of this quadratic equation, and then, depending on whether the roots fall into Case 1, Case 2, or Case 3, write the general solution of the differential equation according to the form given for that Case.

Example 11: Solve the differential equation $y'' + 3y' - 10y = 0$.

The auxiliary polynomial equation is

$$m^2 + 3m - 10 = 0$$

whose roots are real and distinct:

$$m^2 + 3m - 10 = 0$$

$$(m + 5)(m - 2) = 0$$

$$m = -5, 2$$

This problem falls into Case 1, so the general solution of the differential equation is

$$y = c_1 e^{-5x} + c_2 e^{2x} \quad \blacksquare$$

Example 12: Give the general solution of the differential equation $y'' - 2y' + y = 0$.

The auxiliary polynomial equation is

$$m^2 - 2m + 1 = 0$$

which has a double root:

$$m^2 - 2m + 1 = 0$$
$$(m - 1)^2 = 0$$
$$m = 1$$

This problem falls into Case 2, so the general solution of the differential equation is

$$y = c_1e^x + c_2xe^x$$

as you verified earlier in Example 7. ■

Example 13: Solve the differential equation $y'' + 6y' + 25y = 0$.

The auxiliary quadratic equation is

$$m^2 + 6m + 25 = 0$$

which has distinct conjugate complex roots:

$$m = \frac{-6 \pm \sqrt{6^2 - 4 \cdot 1 \cdot 25}}{2 \cdot 1} = \frac{-6 \pm \sqrt{-64}}{2} = \frac{-6 \pm 8i}{2} = -3 \pm 4i$$

This problem falls into Case 3, so the general solution of the differential equation is

$$y = e^{-3x}(c_1 \cos 4x + c_2 \sin 4x) \quad ■$$

Example 14: Solve the IVP

$$2y'' = y - y'$$
$$y(0) = 1$$
$$y'(0) = 2$$

First, rewrite the differential equation in standard form:

$$2y'' + y' - y = 0$$

Next, form the auxiliary polynomial equation

$$2m^2 + m - 1 = 0$$

and determine its roots:

$$2m^2 + m - 1 = 0$$
$$(2m - 1)(m + 1) = 0$$
$$m = \tfrac{1}{2}, -1$$

Since the roots are real and distinct, this problem falls into Case 1, and the general solution of the differential equation is therefore

$$y = c_1 e^{x/2} + c_2 e^{-x} \qquad (*)$$

All that remains is to use the two given initial conditions to determine the values of the constants c_1 and c_2:

$$y(0) = 1 \Rightarrow [c_1 e^{x/2} + c_2 e^{-x}]_{x=0} = 1 \Rightarrow c_1 + c_2 = 1$$
$$y'(0) = 2 \Rightarrow [\tfrac{1}{2} c_1 e^{x/2} - c_2 e^{-x}]_{x=0} = 2 \Rightarrow \tfrac{1}{2} c_1 - c_2 = 2$$

These two equations for c_1 and c_2 can be solved by first adding them to yield

$$\tfrac{3}{2} c_1 = 3 \Rightarrow c_1 = 2$$

then substituting $c_1 = 2$ back into either equation to find $c_2 = -1$.

From (*), the solution of the given IVP is therefore

$$y = 2e^{x/2} - e^{-x} \qquad \blacksquare$$

The Method of Undetermined Coefficients

In order to give the complete solution of a nonhomogeneous linear differential equation, Theorem B says that a particular solution must be added to the general solution of the corresponding homogeneous equation. Now that the method for obtaining the general solution of a homogeneous equation (with constant coefficients) has been discussed, it is time to turn to the problem of determining a particular solution of the original, nonhomogeneous equation.

If the nonhomogeneous term $d(x)$ in the general second-order nonhomogeneous differential equation

$$a(x)y'' + b(x)y' + c(x)y = d(x) \qquad (*)$$

is of a certain special type, then the **method of undetermined coefficients** can be used to obtain a particular solution. The special functions that can be handled by this method are those that have a finite family of derivatives, that is, functions with the property that all their derivatives can be written in terms of just a finite number of other functions.

For example, consider the function $d = \sin x$. Its derivatives are

$$d' = \cos x, \qquad d'' = -\sin x, \qquad d''' = -\cos x, \qquad d^{(\text{iv})} = \sin x,$$

and the cycle repeats. Notice that all derivatives of d can be written in terms of a finite number of functions. [In this case, they are $\sin x$ and $\cos x$, and the set $\{\sin x, \cos x\}$ is called the **family** (of derivatives) of $d = \sin x$.] This is the criterion that describes those nonhomogeneous terms $d(x)$ that make equation (*) susceptible to the method of undetermined coefficients: d must have a finite family.

Here's an example of a function that does not have a finite family of derivatives: $d = \tan x$. Its first four derivatives are

$$d' = \sec^2 x, \qquad d'' = 2\sec^2 x \tan x,$$

$$d''' = 2\sec^4 x + 4\sec^2 x \tan^2 x,$$

$$d^{(iv)} = 16\sec^4 x \tan x + 8\sec^2 x \tan^3 x$$

Notice that the nth derivative ($n \geq 1$) contains a term involving $\tan^{n-1} x$, so as higher and higher derivatives are taken, each one will contain a higher and higher power of $\tan x$, so there is no way that all derivatives can be written in terms of a finite number of functions. The method of undetermined coefficients could not be applied if the nonhomogeneous term in (*) were $d = \tan x$. So just what are the functions $d(x)$ whose derivative families are finite? See Table 3 below.

Table 3
NONZERO FUNCTIONS WITH A FINITE FAMILY OF DERIVATIVES

Function	Family
k (k: a constant)	$\{1\}$
x^n (n: a nonnegative integer)	$\{x^n, x^{n-1}, \ldots, x, 1\}$
e^{kx}	$\{e^{kx}\}$
$\sin kx$	$\{\sin kx, \cos kx\}$
$\cos kx$	$\{\sin kx, \cos kx\}$
a finite product of any of the preceding types	{all products of the individual family members}

Example 15: If $d(x) = 5x^2$, then its family is $\{x^2, x, 1\}$. Note that any numerical coefficients (such as the 5 in this case) are ignored when determining a function's family. ■

Example 16: Since the function $d(x) = x \sin 2x$ is the product of x and $\sin 2x$, the family of $d(x)$ would consist of all products of the family members of the functions x and $\sin 2x$. That is,

$$\begin{aligned}\text{family of } x \sin 2x &= \{x, 1\} \cdot \{\sin 2x, \cos 2x\} \\ &= \{x \sin 2x, x \cos 2x, \sin 2x, \cos 2x\}\end{aligned}$$ ■

Linear combinations of *n* functions. A linear combination of two functions y_1 and y_2 was defined to be any expression of the form

$$c_1y_1 + c_2y_2$$

where c_1 and c_2 are constants. In general, a linear combination of n functions $y_1, y_2, \dots, y_n$ is any expression of the form

$$c_1y_1 + c_2y_2 + \cdots + c_ny_n$$

where $c_1, \dots, c_n$ are constants. Using this terminology, the nonhomogeneous terms $d(x)$ which the method of undetermined coefficients is designed to handle are those for which every derivative can be written as a linear combination of the members of a given finite family of functions.

The central idea of the method of undetermined coefficients is this: Form the most general linear combination of the functions in the family of the nonhomogeneous term $d(x)$, substitute this expression into the given nonhomogeneous differential equation, and solve for the coefficients of the linear combination.

Example 17: Find a particular solution (and the complete solution) of the differential equation

$$y'' + 3y' - 10y = 5x^2$$

As noted in Example 15, the family of $d = 5x^2$ is $\{x^2, x, 1\}$; therefore, the most general linear combination of the functions in the family is $\bar{y} = Ax^2 + Bx + C$ (where A, B, and C are the undetermined coefficients). Substituting this into the given differential equation gives

$$(2A) + 3(2Ax + B) - 10(Ax^2 + Bx + C) = 5x^2$$

Now, combining like terms yields

$$(-10A)x^2 + (6A - 10B)x + (2A + 3B - 10C) = 5x^2$$

In order for this last equation to be an identity, the coefficients of like powers of x on both sides of the equation must be equated. That is, A, B, and C must be chosen so that

$$-10A = 5$$

$$6A - 10B = 0$$

$$2A + 3B - 10C = 0$$

The first equation immediately gives $A = -\frac{1}{2}$. Substituting this into the second equation gives $B = -\frac{3}{10}$, and finally, substituting both of these values into the last equation yields $C = -\frac{19}{100}$. Therefore, a particular solution of the given differential equation is

$$\bar{y} = -\tfrac{1}{2}x^2 - \tfrac{3}{10}x - \tfrac{19}{100}$$

According to Theorem B, then, combining this $\bar{y}$ with the result of Example 11 gives the complete solution of the nonhomogeneous differential equation: $y = c_1e^{-5x} + c_2e^{2x} - \frac{1}{2}x^2 - \frac{3}{10}x - \frac{19}{100}$. ■

Example 18: Find a particular solution (and the complete solution) of the differential equation

$$y'' - 2y' + y = \sin x$$

Since the family of $d = \sin x$ is $\{\sin x, \cos x\}$, the most general linear combination of the functions in the family is $\bar{y} = A \sin x + B \cos x$ (where A and B are the undetermined coefficients). Substituting this into the given differential equation gives

$$(-A \sin x - B \cos x) - 2(A \cos x - B \sin x) + (A \sin x + B \cos x) = \sin x$$

Now, combining like terms and simplifying yields

$$(2B) \sin x + (-2A) \cos x = \sin x$$

In order for this last equation to be an identity, the coefficients A and B must be chosen so that

$$2B = 1$$
$$-2A = 0$$

These equations immediately imply $A = 0$ and $B = \frac{1}{2}$. A particular solution of the given differential equation is therefore

$$\bar{y} = \tfrac{1}{2} \cos x$$

According to Theorem B, combining this $\bar{y}$ with the result of Example 12 yields the complete solution of the given nonhomogeneous differential equation: $y = c_1 e^x + c_2 x e^x + \frac{1}{2} \cos x$. ■

Example 19: Find a particular solution (and the complete solution) of the differential equation

$$y'' + 6y' + 25y = 8e^{-7x}$$

Since the family of $d = 8e^{-7x}$ is just $\{e^{-7x}\}$, the most general linear combination of the functions in the family is simply $\bar{y} = Ae^{-7x}$ (where A is the undetermined coefficient). Substituting this into the given differential equation gives

$$(49Ae^{-7x}) + 6(-7Ae^{-7x}) + 25(Ae^{-7x}) = 8e^{-7x}$$

Simplifying yields

$$32Ae^{-7x} = 8e^{-7x}$$

In order for this last equation to be an identity, the coefficient A must be chosen so that

$$32A = 8$$

which immediately gives $A = \frac{1}{4}$. A particular solution of the given differential equation is therefore

$$\bar{y} = \tfrac{1}{4}e^{-7x}$$

and then, according to Theorem B, combining $\bar{y}$ with the result of Example 13 gives the complete solution of the nonhomogeneous differential equation: $y = e^{-3x}(c_1 \cos 4x + c_2 \sin 4x) + \frac{1}{4}e^{-7x}$. ■

Example 20: Find the solution of the IVP

$$y'' - y' - 6y = -e^x + 12x$$

$$y(0) = 1$$

$$y'(0) = -2$$

The first step is to obtain the general solution of the corresponding homogeneous equation

$$y'' - y' - 6y = 0$$

Since the auxiliary polynomial equation has distinct real roots,

$$m^2 - m - 6 = 0$$
$$(m + 2)(m - 3) = 0$$
$$m = -2, 3$$

the general solution of the corresponding homogeneous equation is $y_h = c_1 e^{-2x} + c_2 e^{3x}$.

Now, since the nonhomogeneous term $d(x)$ is a (finite) sum of functions from Table 3, the family of $d(x)$ is the *union* of the families of the individual functions. That is, since the family of $-e^x$ is $\{e^x\}$, and the family of $12x$ is $\{x, 1\}$,

$$\text{the family of } -e^x + 12x = [\text{the family of } -e^x] \cup [\text{the family of } 12x] = \{e^x\} \cup \{x, 1\} = \{e^x, x, 1\}$$

The most general linear combination of the functions in the family of $d = -e^x + 12x$ is therefore $\bar{y} = Ae^x + Bx + C$ (where A, B, and C are the undetermined coefficients). Substituting this into the given differential equation gives

$$(Ae^x) - (Ae^x + B) - 6(Ae^x + Bx + C) = -e^x + 12x$$

Combining like terms and simplifying yields

$$(-6A)e^x + (-6B)x + (-B - 6C) = -e^x + 12x$$

In order for this last equation to be an identity, the coefficients A, B, and C must be chosen so that

$$-6A = -1$$

$$-6B = 12$$

$$-B - 6C = 0$$

The first two equations immediately give $A = \frac{1}{6}$ and $B = -2$, whereupon the third implies $C = \frac{1}{3}$. A particular solution of the given differential equation is therefore

$$\bar{y} = \tfrac{1}{6}e^x - 2x + \tfrac{1}{3}$$

According to Theorem B, then, combining this $\bar{y}$ with y_h gives the complete solution of the nonhomogeneous differential equation: $y = c_1e^{-2x} + c_2e^{3x} + \frac{1}{6}e^x - 2x + \frac{1}{3}$. Now, to apply the initial conditions and evaluate the parameters c_1 and c_2:

$$y(0) = 1 \Rightarrow [c_1e^{-2x} + c_2e^{3x} + \tfrac{1}{6}e^x - 2x + \tfrac{1}{3}]_{x=0} = 1 \Rightarrow c_1 + c_2 + \tfrac{1}{2} = 1$$

$$y'(0) = -2 \Rightarrow [-2c_1e^{-2x} + 3c_2e^{3x} + \tfrac{1}{6}e^x - 2]_{x=0} = -2$$

$$\Rightarrow -2c_1 + 3c_2 - \tfrac{11}{6} = -2$$

Solving these last two equations yields $c_1 = \frac{1}{3}$ and $c_2 = \frac{1}{6}$. Therefore, the desired solution of the IVP is

$$y = \tfrac{1}{3}(e^{-2x} + 1) + \tfrac{1}{6}(e^x + e^{3x}) - 2x \quad \blacksquare$$

Now that the basic process of the method of undetermined coefficients has been illustrated, it is time to mention that it isn't always this straightforward. *A problem arises if a member of a family of the nonhomogeneous term happens to be a solution of the corresponding homogeneous equation*. In this case, that family must be modified before the general linear combination can be substituted into the original nonhomogeneous differential equation to solve for the undetermined coefficients. The specific modification procedure will be introduced through the following alteration of Example 20.

Example 21: Find the complete solution of the differential equation

$$y'' - y' - 6y = 10e^{3x}$$

The general solution of the corresponding homogeneous equation was obtained in Example 20:

$$y_h = c_1e^{-2x} + c_2e^{3x}$$

Note carefully that the family $\{e^{3x}\}$ of the nonhomogeneous term $d = 10e^{3x}$ contains a solution of the corresponding homogeneous equation (take $c_1 = 0$ and $c_2 = 1$ in the expression for y_h). The "offending" family is modified as follows: *Multiply each member of the family by x and try again.*

$$\{e^{3x}\} \xrightarrow[\text{by } x]{\text{multiply each family member}} \{xe^{3x}\}$$

Since the modified family no longer contains a solution of the corresponding homogeneous equation, the method of undetermined coefficients can now proceed. (If xe^{3x} had been again a solution of the corresponding homogeneous equation, you would perform the modification procedure once more: *Multiply each member of the family by x and try again.*) Therefore, substituting $\bar{y} = Axe^{3x}$ into the given nonhomogeneous differential equation yields

$$(9Axe^{3x} + 6Ae^{3x}) - (3Axe^{3x} + Ae^{3x}) - 6(Axe^{3x}) = 10e^{3x}$$

$$5Ae^{3x} = 10e^{3x}$$

$$A = 2$$

This calculation implies that $\bar{y} = 2xe^{3x}$ is a particular solution of the nonhomogeneous equation, so combining this with y_h gives the complete solution:

$$y = c_1e^{-2x} + c_2e^{3x} + 2xe^{3x} \quad ■$$

Example 22: Find the complete solution of the differential equation

$$y'' - 2y' = 6x^2 - 3e^{x/2}$$

First, obtain the general solution of the corresponding homogeneous equation

$$y'' - 2y' = 0$$

Since the auxiliary polynomial equation has distinct real roots,

$$m^2 - 2m = 0$$

$$m(m - 2) = 0$$

$$m = 0, 2$$

the general solution of the corresponding homogeneous equation is

$$y_h = c_1e^{0x} + c_2e^{2x} = c_1 + c_2e^{2x}$$

The family for the $6x^2$ term is $\{x^2, x, 1\}$, and the family for the $-3e^{x/2}$ term is simply $\{e^{x/2}\}$. This latter family does not contain a solution of the corresponding homogeneous equation, but the family $\{x^2, x, 1\}$ *does* (it contains the constant function 1, which matches y_h when $c_1 = 1$ and $c_2 = 0$). This entire family (not just the "offending" member) must therefore be modified:

$$\{x^2, x, 1\} \xrightarrow{\text{multiply each family member by } x} \{x^3, x^2, x\}$$

The family that will be used to construct the linear combination $\bar{y}$ is now the union

$$\{x^3, x^2, x\} \cup \{e^{x/2}\} = \{x^3, x^2, x, e^{x/2}\}$$

This implies that $\bar{y} = Ax^3 + Bx^2 + Cx + De^{x/2}$ (where A, B, C, and D are the undetermined coefficients) should be substituted into the given nonhomogeneous differential equation. Doing so yields

$$(6Ax + 2B + \tfrac{1}{4}De^{x/2}) - 2(3Ax^2 + 2Bx + C + \tfrac{1}{2}De^{x/2}) = 6x^2 - 3e^{x/2}$$

which after combining like terms reads

$$(-6A)x^2 + (6A - 4B)x + (2B - 2C) + (-\tfrac{3}{4}De^{x/2}) = 6x^2 - 3e^{x/2}$$

In order for this last equation to be an identity, the coefficients A, B, C, and D must be chosen so that

$$-6A = 6$$

$$6A - 4B = 0$$

$$2B - 2C = 0$$

$$-\tfrac{3}{4}D = -3$$

These equations determine the values of the coefficients: $A = -1$, $B = C = -\frac{3}{2}$, and $D = 4$. Therefore, a particular solution of the given differential equation is

$$\bar{y} = -x^3 - \tfrac{3}{2}x^2 - \tfrac{3}{2}x + 4e^{x/2}$$

According to Theorem B, then, combining this $\bar{y}$ with y_h gives the complete solution of the nonhomogeneous differential equation: $y = c_1 + c_2e^{2x} - x^3 - \frac{3}{2}x^2 - \frac{3}{2}x + 4e^{x/2}$. ■

Example 23: Find the complete solution of the equation

$$y'' + 4y = x \sin 2x + 8$$

First, obtain the general solution of the corresponding homogeneous equation

$$y'' + 4y = 0$$

Since the auxiliary polynomial equation has distinct conjugate complex roots,

$$m^2 + 4 = 0$$
$$m = \pm 2i$$

the general solution of the corresponding homogeneous equation is

$$y_h = c_1 \cos 2x + c_2 \sin 2x$$

Example 16 above showed that the

$$\text{family of } x \sin 2x = \{x, 1\} \cdot \{\sin 2x, \cos 2x\}$$
$$= \{x \sin 2x, x \cos 2x, \sin 2x, \cos 2x\}$$

Note that this family contains $\sin 2x$ and $\cos 2x$, which are solutions of the corresponding homogeneous equation. Therefore, this entire family must be modified:

$$\{x \sin 2x, x \cos 2x, \sin 2x, \cos 2x\} \xrightarrow{\text{multiply each family member by } x}$$
$$\{x^2 \sin 2x, x^2 \cos 2x, x \sin 2x, x \cos 2x\}$$

None of the members of this family are solutions of the corresponding homogeneous equation, so the solution can now proceed as usual. Since the family of the constant term is simply $\{1\}$, the family used to construct $\bar{y}$ is the union

$$\{x^2 \sin 2x, x^2 \cos 2x, x \sin 2x, x \cos 2x\} \cup \{1\}$$
$$= \{x^2 \sin 2x, x^2 \cos 2x, x \sin 2x, x \cos 2x, 1\}$$

This implies that $\bar{y} = Ax^2 \sin 2x + Bx^2 \cos 2x + Cx \sin 2x + Dx \cos 2x + E$ (where A, B, C, D, and E are the undetermined coefficients) should be substituted into the given nonhomogeneous differential equation $y'' + 4y = x \sin 2x + 8$. Doing so yields

$$[-8Bx + (2A - 4D)] \sin 2x + [8Ax + (2B + 4C)] \cos 2x + 4E = x \sin 2x + 8$$

In order for this last equation to be an identity, A, B, C, D, and E must be chosen so that

$$-8B = 1$$
$$2A - 4D = 0$$
$$8A = 0$$
$$2B + 4C = 0$$
$$4E = 8$$

These equations determine the coefficients: $A = 0$, $B = -\frac{1}{8}$, $C = \frac{1}{16}$, $D = 0$, and $E = 2$. Therefore, a particular solution of the given differential equation is

$$\bar{y} = -\tfrac{1}{8}x^2 \cos 2x + \tfrac{1}{16}x \sin 2x + 2$$

According to Theorem B, then, combining this $\bar{y}$ with y_h gives the complete solution of the nonhomogeneous differential equation: $y = c_1 \cos 2x + c_2 \sin 2x - \frac{1}{8}x^2 \cos 2x + \frac{1}{16}x \sin 2x + 2$. ■

Variation of Parameters

For the differential equation

$$a(x)y'' + b(x)y' + c(x)y = d(x) \quad (*)$$

the method of undetermined coefficients works only when the coefficients a, b, and c are constants and the right-hand term $d(x)$ is of a special form. If these restrictions do not apply to a given nonhomogeneous linear differential equation, then a more powerful method of determining a particular solution is needed: the method known as **variation of parameters.**

The first step is to obtain the general solution of the corresponding homogeneous equation, which will have the form

$$y_h = c_1y_1 + c_2y_2$$

where y_1 and y_2 are known functions. The next step is to *vary the parameters*; that is, to replace the constants c_1 and c_2 by (as yet unknown) functions $v_1(x)$ and $v_2(x)$ to obtain the form of a particular solution $\bar{y}$ of the given nonhomogeneous equation:

$$\bar{y} = v_1(x)y_1 + v_2(x)y_2$$

The goal is to determine these functions v_1 and v_2. Then, since the functions y_1 and y_2 are already known, the expression above for $\bar{y}$ yields a particular solution of the nonhomogeneous equation. Combining $\bar{y}$ with y_h then gives the general solution of the nonhomogeneous differential equation, as guaranteed by Theorem B.

Since there are two unknowns to be determined, v_1 and v_2, two equations or conditions are required to obtain a solution. One of these conditions will naturally be satisfying the given differential equation. But another condition will be imposed first. Since $\bar{y}$ will be substituted into equation (*), its derivatives must be evaluated. The first derivative of $\bar{y}$ is

$$\bar{y}' = v_1y_1' + v_1'y_1 + v_2y_2' + v_2'y_2$$

Now, to simplify the rest of the process—and to produce the first condition on v_1 and v_2—set

$$v_1'y_1 + v_2'y_2 = 0$$

This will always be the first condition in determining v_1 and v_2; the second condition will be the satisfaction of the given differential equation (*).

Example 24: Give the general solution of the differential equation $y'' + y = \tan x$.

Since the nonhomogeneous right-hand term, $d = \tan x$, is not of the special form that the method of undetermined coefficients can handle, variation of parameters is required. The first step is to obtain the general solution of the corresponding homogeneous equation, $y'' + y = 0$. The auxiliary polynomial equation is

$$m^2 + 1 = 0$$

whose roots are the distinct conjugate complex numbers $m = \pm i = 0 \pm 1i$. The general solution of the homogeneous equation is therefore

$$y = c_1 \sin x + c_2 \cos x$$

Now, vary the parameters c_1 and c_2 to obtain

$$\bar{y} = v_1 \sin x + v_2 \cos x$$

Differentiation yields

$$\bar{y}' = v_1 \cos x + v_1' \sin x - v_2 \sin x + v_2' \cos x$$

Next, remember the first condition to be imposed on v_1 and v_2:

$$v_1'y_1 + v_2'y_2 = 0$$

that is,

$$v_1' \sin x + v_2' \cos x = 0 \quad (1)$$

This reduces the expression for $\bar{y}'$ to

$$\bar{y}' = v_1 \cos x - v_2 \sin x$$

so, then,

$$\bar{y}'' = -v_1 \sin x + v_1' \cos x - v_2 \cos x - v_2' \sin x$$

Substitution into the given nonhomogeneous equation $y'' + y = \tan x$ yields

$$(-v_1 \sin x + v_1' \cos x - v_2 \cos x - v_2' \sin x) + (v_1 \sin x + v_2 \cos x) = \tan x$$

$$v_1' \cos x - v_2' \sin x = \tan x \quad (2)$$

Therefore, the two conditions on v_1 and v_2 are

$$v_1' \sin x + v_2' \cos x = 0 \quad (1)$$

$$v_1' \cos x - v_2' \sin x = \tan x \quad (2)$$

To solve these two equations for v_1' and v_2', first multiply the first equation by $\sin x$; then multiply the second equation by $\cos x$:

$$v_1' \sin^2 x + v_2' \cos x \sin x = 0$$

$$v_1' \cos^2 x - v_2' \sin x \cos x = \sin x$$

Adding these equations yields

$$v_1'(\sin^2 x + \cos^2 x) = \sin x \Rightarrow v_1' = \sin x$$

Substituting $v_1' = \sin x$ back into equation (1) [or equation (2)] then gives

$$v_2' = -\tan x \sin x$$

Now, integrate to find v_1 and v_2 (and ignore the constant of integration in each case):

$$\begin{aligned} v_1 &= \int v_1' \\ &= \int \sin x \, dx \\ &= -\cos x \end{aligned}$$

and

$$\begin{aligned} v_2 &= \int v_2' \\ &= \int -\tan x \sin x \, dx \\ &= \int \frac{-\sin^2 x}{\cos x} \, dx \\ &= \int \frac{\cos^2 x - 1}{\cos x} \, dx \\ &= \int (\cos x - \sec x) \, dx \\ &= \sin x - \ln|\sec x + \tan x| \end{aligned}$$

Therefore, a particular solution of the given nonhomogeneous differential equation is

$$\begin{aligned} \bar{y} &= v_1 \sin x + v_2 \cos x \\ &= -\cos x \sin x + [\sin x - \ln|\sec x + \tan x|] \cos x \\ \bar{y} &= -\cos x \ln|\sec x + \tan x| \end{aligned}$$

Combining this with the general solution of the corresponding homogeneous equation gives the general solution of the nonhomogeneous equation:

$$y = c_1 \sin x + c_2 \cos x - \cos x \ln|\sec x + \tan x| \quad ■$$

In general, when the method of variation of parameters is applied to the second-order nonhomogeneous linear differential equation

$$a(x)y'' + b(x)y' + c(x)y = d(x)$$

with $\bar{y} = v_1(x)y_1 + v_2(x)y_2$ (where $y_h = c_1y_1 + c_2y_2$ is the general solution of the corresponding homogeneous equation), the two conditions on v_1 and v_2 will always be

$$v_1'y_1 + v_2'y_2 = 0 \qquad (1)$$

$$a(x)[v_1'y_1' + v_2'y_2'] = d(x) \qquad (2)$$

So after obtaining the general solution of the corresponding homogeneous equation ($y_h = c_1y_1 + c_2y_2$) and varying the parameters by writing $\bar{y} = v_1y_1 + v_2y_2$, go directly to equations (1) and (2) above and solve for v_1' and v_2'.

Example 25: Give the general solution of the differential equation

$$y'' - 2y' + y = e^x \ln x$$

Because of the $\ln x$ term, the right-hand side is not one of the special forms that the method of undetermined coefficients can handle; variation of parameters is required. The first step, obtaining the general solution of the corresponding homogeneous equation, $y'' - 2y' + y = 0$, was done in Example 12:

$$y_h = c_1y_1 + c_2y_2 = c_1e^x + c_2xe^x$$

Varying the parameters gives the particular solution

$$\bar{y} = v_1y_1 + v_2y_2 = v_1e^x + v_2xe^x$$

and the system of equations (1) and (2) becomes

$$v_1'e^x + v_2'xe^x = 0 \qquad (1)$$

$$v_1'e^x + v_2'e^x(x + 1) = e^x \ln x \qquad (2)$$

Cancel out the common factor of e^x in both equations; then subtract the resulting equations to obtain

$$v_2' = \ln x$$

Substituting this back into either equation (1) or (2) determines

$$v_1' = -x \ln x$$

Now, integrate (by parts, in both these cases) to obtain v_1 and v_2 from v_1' and v_2':

$$v_1 = \int v_1' = -\int x \ln x \, dx = -\left[(\ln x)\left(\frac{1}{2}x^2\right) - \int \left(\frac{1}{2}x^2\right)\left(\frac{1}{x}\,dx\right)\right]$$

$$= \tfrac{1}{4}x^2 (1 - 2 \ln x)$$

$$v_2 = \int v_2' = \int \ln x \, dx = (\ln x)(x) - \int (x)\left(\frac{1}{x}\,dx\right) = x(\ln x - 1)$$

Therefore, a particular solution is

$$\bar{y} = v_1y_1 + v_2y_2$$

$$= v_1e^x + v_2xe^x$$

$$= \tfrac{1}{4}x^2(1 - 2 \ln x)e^x + x(\ln x - 1)xe^x$$

$$\bar{y} = \tfrac{1}{4}x^2e^x (2 \ln x - 3)$$

Consequently, the general solution of the given nonhomogeneous equation is

$$y = c_1e^x + c_2xe^x + \tfrac{1}{4}x^2e^x (2 \ln x - 3) \quad \blacksquare$$

Example 26: Give the general solution of the following differential equation, given that $y_1 = x$ and $y_2 = x^3$ are solutions of its corresponding homogeneous equation:

$$x^2y'' - 3xy' + 3y = 12x^4$$

Since the functions $y_1 = x$ and $y_2 = x^3$ are linearly independent, Theorem A says that the general solution of the corresponding homogeneous equation is

$$y_h = c_1y_1 + c_2y_2 = c_1x + c_2x^3$$

Varying the parameters c_1 and c_2 gives the form of a particular solution of the given nonhomogeneous equation:

$$\bar{y} = v_1y_1 + v_2y_2 = v_1x + v_2x^3$$

where the functions v_1 and v_2 are as yet undetermined. The two conditions on v_1 and v_2 which follow from the method of variation of parameters are

$$v_1'y_1 + v_2'y_2 = 0 \qquad (1)$$

$$a(x)[v_1'y_1' + v_2'y_2'] = d(x) \qquad (2)$$

which in this case ($y_1 = x, y_2 = x^3, a = x^2, d = 12x^4$) become

$$v_1'x + v_2'x^3 = 0 \qquad (1)$$

$$x^2[v_1' + v_2' \cdot 3x^2] = 12x^4 \qquad (2)$$

Solving this system for v_1' and v_2' yields

$$v_1' = -6x^2 \quad \text{and} \quad v_2' = 6$$

from which follow

$$v_1 = \int v_1' = \int (-6x^2)\,dx = -2x^3$$

$$v_2 = \int v_2' = \int 6\,dx = 6x$$

Therefore, the particular solution obtained is

$$\begin{aligned}\bar{y} &= v_1y_1 + v_2y_2 \\ &= v_1x + v_2x^3 \\ &= (-2x^3)x + (6x)x^3 \\ \bar{y} &= 4x^4\end{aligned}$$

and the general solution of the given nonhomogeneous equation is

$$y = c_1x + c_2x^3 + 4x^4 \qquad \blacksquare$$

The Cauchy-Euler Equidimensional Equation

The second-order homogeneous **Cauchy-Euler equidimensional equation** has the form

$$ax^2y'' + bxy' + cy = 0$$

where a, b, and c are constants (and $a \neq 0$). The quickest way to solve this linear equation is to substitute $y = x^m$ and solve for m. If $y = x^m$, then

$$y' = mx^{m-1} \quad \text{and} \quad y'' = m(m-1)x^{m-2}$$

so substitution into the differential equation yields

$$\begin{aligned}am(m-1)x^m + bmx^m + cx^m &= 0 \\ x^m[am(m-1) + bm + c] &= 0 \\ am(m-1) + bm + c &= 0 \\ am^2 + (b-a)m + c &= 0 \qquad (*)\end{aligned}$$

Just as in the case of solving second-order linear homogeneous equations with constant coefficients (by first setting $y = e^{mx}$ and then solving the resulting auxiliary quadratic equation for m), this process of solving the equidimensional equation also yields an auxiliary quadratic polynomial equation. The question here is, how is $y = x^m$ to be interpreted to give two linearly independent solutions (and thus the general solution) in each of the three cases for the roots of the resulting quadratic equation?

Case 1: The roots of (*) are real and distinct.

If the two roots are denoted m_1 and m_2, then the general solution of the second-order homogeneous equidimensional differential equation in this case is

$$y = c_1x^{m_1} + c_2x^{m_2}$$

Case 2: The roots of (*) are real and identical.

If the double (repeated) root is denoted simply by m, then the general solution (for $x > 0$) of the homogeneous equidimensional differential equation in this case is

$$y = c_1x^m + c_2x^m \ln x$$

Case 3: The roots of (*) are distinct conjugate complex numbers.

If the roots are denoted $r \pm si$, then the general solution of the homogeneous equidimensional differential equation in this case is

$$y = x^r[c_1 \cos(s \ln x) + c_2 \sin(s \ln x)]$$

Example 27: Give the general solution of the equidimensional equation

$$x^2y'' - 3xy' + 3y = 0$$

Substitution of $y = x^m$ results in

$$x^2 \cdot m(m-1)x^{m-2} - 3x \cdot mx^{m-1} + 3x^m = 0$$
$$x^m[m(m-1) - 3m + 3] = 0$$
$$m(m-1) - 3m + 3 = 0$$
$$m^2 - 4m + 3 = 0$$
$$(m-1)(m-3) = 0$$
$$m = 1, 3$$

Since the roots of the resulting quadratic equation are real and distinct (Case 1), both $y = x^1 = x$ and $y = x^3$ are solutions and linearly independent (see Example 17), and the general solution of this homogeneous equation is

$$y = c_1x + c_2x^3$$

(Compare this result with the statement of Example 26.) ■

Example 28: For the following equidimensional equation, give the general solution which is valid in the domain $x > 0$:

$$x^2y'' - 3xy' + 4y = 0$$

Substitution of $y = x^m$ results in

$$x^2 \cdot m(m-1)x^{m-2} - 3x \cdot mx^{m-1} + 4x^m = 0$$
$$x^m[m(m-1) - 3m + 4] = 0$$
$$m(m-1) - 3m + 4 = 0$$
$$m^2 - 4m + 4 = 0$$
$$(m-2)^2 = 0$$
$$m = 2, 2$$

Since the roots of the resulting quadratic equation are real and identical (Case 2), both $y = x^2$ and $y = x^2 \ln x$ are (linearly independent) solutions, so the general solution (valid for $x > 0$) of this homogeneous equation is

$$y = c_1x^2 + c_2x^2 \ln x \quad \blacksquare$$

If the general solution of a *non*homogeneous equidimensional equation is desired, first use the method above to obtain the general solution of the corresponding homogeneous equation; then apply variation of parameters (see Example 26).

Reduction of Order

Some second-order equations can be reduced to first-order equations, rendering them susceptible to the simple methods of solving equations of the first order. Three particular types of such second-order equations will be discussed in this section.

Type 1: Second-order equations with the dependent variable missing

Type 2: Second-order nonlinear equations with the independent variable missing

Type 3: Second-order homogeneous linear equations where one (nonzero) solution is known

Type 1: Second-order equations with the dependent variable missing. Examples of such equations include

$$y'' + y' = x \quad \text{and} \quad xy'' - 2y' = 12x^2$$

The defining characteristic is this: The dependent variable, y, does not explicitly appear in the equation. This type of second-order equation is easily reduced to a first-order equation by the transformation

$$y' = w$$

This substitution obviously implies $y'' = w'$, and the original equation becomes a first-order equation for w. Solve for the function w; then integrate it to recover y.

Example 29: Solve the differential equation $y'' + y' = x$.

Since the dependent variable y is missing, let $y' = w$ and $y'' = w'$. These substitutions transform the given second-order equation into the first-order equation

$$w' + w = x$$

which is in standard form. Applying the method for solving such equations, the integrating factor is first determined,

$$\mu = e^{\int P\,dx} = e^{\int dx} = e^x$$

and then used to multiply both sides of the equation, yielding

$$e^x w' + e^x w = xe^x$$

$$\frac{d}{dx}(e^x w) = xe^x$$

$$\begin{aligned} e^x w &= \int xe^x\,dx \\ &= xe^x - \int e^x\,dx \\ &= xe^x - (e^x + c_1) \end{aligned}$$

Therefore,

$$w = x - 1 - c_1 e^{-x}$$

Now, to give the solution y of the original second-order equation, integrate:

$$y' = w \Rightarrow y = \int w = \int (x - 1 - c_1e^{-x})\,dx$$

This gives

$$y = \tfrac{1}{2}x^2 - x + c_1e^{-x} + c_2$$

Referring to Theorem B, note that this solution implies that $y = c_1e^{-x} + c_2$ is the general solution of the corresponding homogeneous equation and that $\bar{y} = \frac{1}{2}x^2 - x$ is a particular solution of the nonhomogeneous equation. (This particular differential equation could also have been solved by applying the method for solving second-order linear equations with constant coefficients.) ■

Example 30: Solve the differential equation

$$xy'' - 2y' = 10x^4$$

Again, the dependent variable y is missing from this second-order equation, so its order will be reduced by making the substitutions $y' = w$ and $y'' = w'$:

$$xw' - 2w = 10x^4$$

which can be written in standard form

$$w' - \frac{2}{x}w = 10x^3$$

The integrating factor here is

$$\mu = e^{\int P\,dx} = e^{\int(-2/x)dx} = e^{-2\ln x} = e^{\ln(x^{-2})} = x^{-2}$$

which is used to multiply both sides of the equation, yielding

$$x^{-2}w' - 2x^{-3}w = 10x$$

$$\frac{d}{dx}(x^{-2}w) = 10x$$

$$x^{-2}w = \int 10x\,dx$$

$$x^{-2}w = 5x^2 + c_1'$$

$$w = 5x^4 + c_1'x^2$$

Integrating w gives y:

$$y' = w \Rightarrow y = \int w$$

$$= \int (5x^4 + c_1'x^2)\,dx$$

$$= x^5 + \tfrac{1}{3}c_1'x^3 + c_2$$

Letting $c_1 = \frac{1}{3}c_1'$, the general solution can be written

$$y = x^5 + c_1x^3 + c_2 \qquad \blacksquare$$

Example 31: Sketch the solution of the IVP

$$y'' + (y')^2 = 0$$

$$y(0) = 1$$

$$y'(0) = 1$$

Although this equation is nonlinear [because of the term $(y')^2$; neither y nor any of its derivatives are allowed to be raised to any power (other than 1) in a linear equation], the substitutions $y' = w$ and $y'' = w'$ will still reduce this to a first-order equation, since the variable y does not explicitly appear. The differential equation is transformed into

$$w' + w^2 = 0$$

which is separable:

$$\frac{dw}{dx} = -w^2$$

$$-\frac{dw}{w^2} = dx$$

$$\int -\frac{dw}{w^2} = \int dx$$

$$\frac{1}{w} = x + c_1$$

$$w = \frac{1}{x + c_1}$$

Since $y' = w$, integrating gives

$$y = \int w = \int \frac{1}{x + c_1} dx = \ln(x + c_1) + c_2$$

Now apply the initial conditions to determine the constants c_1 and c_2:

$$y(0) = 1 \Rightarrow [\ln(x + c_1) + c_2]_{x=0} = 1 \Rightarrow \ln c_1 + c_2 = 1$$

$$y'(0) = 1 \Rightarrow [(x + c_1)^{-1}]_{x=0} = 1 \Rightarrow c_1 = 1$$

Because $c_1 = 1$, the first condition then implies $c_2 = 1$ also. Thus the solution of this IVP (at least for $x > -1$) is

$$y = \ln(x + 1) + 1$$

whose graph is shown in Figure 7.

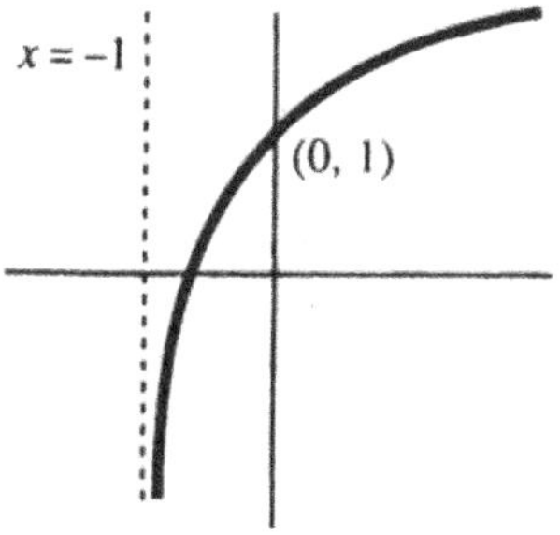

■ Figure 7 ■

Type 2: Second-order nonlinear equations with the independent variable missing. Here's an example of such an equation:

$$y^2y'' - (y')^3 = 0$$

The defining characteristic is this: The independent variable, x, does not explicitly appear in the equation.

The method for reducing the order of these second-order equations begins with the same substitution as for Type 1 equations, namely, replacing y' by w. But instead of simply writing y'' as w', the trick here is to express y'' in terms of a first derivative *with respect to* y. This is accomplished using the chain rule:

$$y'' = \frac{d}{dx}\left(\frac{dy}{dx}\right) = \frac{dw}{dx} = \frac{dw}{dy}\frac{dy}{dx} = \frac{dw}{dy}y'$$

Therefore,

$$y'' = w\frac{dw}{dy}$$

This substitution, along with $y' = w$, will reduce a Type 2 equation to a first-order equation for w. Once w is determined, integrate to find y.

Example 32: Solve the differential equation

$$y^2y'' - (y')^3 = 0$$

The substitutions $y' = w$ and $y'' = w(dw/dy)$ transform this second-order equation for y into the following first-order equation for w:

$$y^2w\frac{dw}{dy} - w^3 = 0$$

$$w\left(y^2\frac{dw}{dy} - w^2\right) = 0$$

Therefore,

$$w = 0 \quad \text{or} \quad y^2\frac{dw}{dy} - w^2 = 0$$

The statement $w = 0$ means $y' = 0$, and thus $y = c$ is a solution for any constant c. The second statement is a separable equation, and its solution proceeds as follows:

$$y^2\frac{dw}{dy} = w^2$$

$$\frac{dw}{w^2} = \frac{dy}{y^2}$$

$$\int\frac{dw}{w^2} = \int\frac{dy}{y^2}$$

$$-\frac{1}{w} = -\frac{1}{y} - c_1$$

$$\frac{1}{w} = \frac{1}{y} + c_1$$

Now, since $w = dy/dx$, this last result becomes

$$\frac{dx}{dy} = \frac{1}{y} + c_1$$

$$(y^{-1} + c_1)\, dy = dx$$

$$\int (y^{-1} + c_1)\, dy = \int dx$$

which gives the general solution, expressed implicitly as follows:

$$\ln|y| + c_1 y = x + c_2$$

Therefore, the complete solution of the given differential equation is

$$y = c \quad \text{or} \quad \ln|y| + c_1 y = x + c_2 \qquad \blacksquare$$

Type 3: Second-order homogeneous linear equations where one (nonzero) solution is known. Sometimes it is possible to determine a solution of a second-order differential equation by inspection, which usually amounts to successful trial and error with a few particularly simple functions. For example, you might discover that the simple function $y = x$ is a solution of the equation

$$x^2y'' - xy' + y = 0$$

or that $y = e^x$ satisfies the equation

$$xy'' - (x + 1)y' + y = 0$$

Of course, trial and error is not the best way to solve an equation, but if you are lucky (or practiced) enough to actually discover a solution by inspection, you should be rewarded.

If one (nonzero) solution of a homogeneous second-order equation is known, there is a straightforward process for determining a second, linearly independent solution, which can then be combined with the first one to give the general solution. Let y_1 denote the function you know is a solution. Then let $y = y_1v(x)$, where v is a function (as yet unknown). Substitute $y = y_1v$ into the differential equation and derive a second-order equation for v. This will turn out to be a Type 1 equation for v (because the dependent variable, v, will not explicitly appear). Use the technique described earlier to solve for the function v; then substitute into the expression $y = y_1v$ to give the desired second solution.

Example 33: Give the general solution of the differential equation

$$x^2y'' - xy' + y = 0$$

As mentioned above, it is easy to discover the simple solution $y = x$. Denoting this known solution by y_1, substitute $y = y_1v = xv$ into the given differential equation and solve for v. If $y = xv$, then the derivatives are

$$y' = xv' + v$$

$$y'' = xv'' + 2v'$$

Substitution into the differential equation yields

$$x^2(xv'' + 2v') - x(xv' + v) + xv = 0$$

$$x^3v'' + x^2v' = 0$$

$$v'' + \frac{1}{x}v' = 0$$

Note that this resulting equation is a Type 1 equation for v (because the dependent variable, v, does not explicitly appear). So, letting $v' = w$ and $v'' = w'$, this second-order equation for v becomes the following first-order equation for w:

$$w' + \frac{1}{x}w = 0 \qquad (*)$$

The integrating factor for this standard first-order linear equation is

$$\mu = e^{\int P\,dx} = e^{\int (1/x)\,dx} = e^{\ln x} = x$$

and multiplying both sides of (*) by $\mu = x$ gives

$$xw' + w = 0$$

$$\frac{d}{dx}(xw) = 0$$

$$xw = c$$

$$w = cx^{-1}$$

Ignore the constant c and integrate to recover v:

$$v' = w \Rightarrow v = \int w = \int x^{-1}\,dx = \ln|x|$$

Multiply this by y_1 to obtain the desired second solution,

$$y_2 = y_1 v = x \ln|x|$$

The general solution of the original equation is any linear combination of $y_1 = x$ and $y_2 = x \ln|x|$:

$$y = c_1 x + c_2 x \ln|x|$$

This agrees with the general solution that would be found if this problem were attacked using the method for solving an equidimensional equation. ■

Example 34: Determine the general solution of the following differential equation, given that it is satisfied by the function $y = e^x$:

$$xy'' - (x + 1)y' + y = 0$$

Denoting the known solution by y_1, substitute $y = y_1 v = e^x v$ into the differential equation. With $y = e^x v$, the derivatives are

$$y' = e^x v' + e^x v$$

$$y'' = e^x v'' + 2e^x v' + e^x v$$

Substitution into the given differential equation yields

$$x(e^x v'' + 2e^x v' + e^x v) - (x + 1)(e^x v' + e^x v) + e^x v = 0$$

which simplifies to the following Type 1 second-order equation for v:

$$xv'' + (x - 1)v' = 0$$

Letting $v' = w$, then rewriting the equation in standard form, yields

$$w' + \frac{x - 1}{x} w = 0 \qquad (*)$$

The integrating factor in this case is

$$\mu = e^{\int P\,dx} = e^{\int [1 - (1/x)]\,dx} = e^{x - \ln x} = e^x e^{-\ln x} = \frac{e^x}{x}$$

Multiplying both sides of (*) by $\mu = e^x/x$ yields

$$\frac{d}{dx}\left(\frac{e^x}{x} w\right) = 0$$

$$\frac{e^x}{x} w = c$$

$$w = cxe^{-x}$$

Ignore the constant c and integrate to recover v:

$$\begin{aligned} v &= \int w \\ &= \int xe^{-x}\,dx \\ &= -xe^{-x} - \int -e^{-x}\,dx \\ &= -xe^{-x} - e^{-x} \\ &= -e^{-x}(x+1) \end{aligned}$$

Multiply this by y_1 to obtain the desired second solution,

$$y_2 = y_1 v = e^x \cdot [-e^{-x}(x+1)] = -(x+1)$$

The general solution of the original equation is any linear combination of y_1 and y_2:

$$y = c_1 e^x + c_2(x+1) \qquad \blacksquare$$

It often happens that a differential equation cannot be solved in terms of **elementary** functions (that is, in closed form in terms of polynomials, rational functions, e^x, $\sin x$, $\cos x$, $\ln x$, etc.). A power series solution is all that is available. Such an expression is nevertheless an entirely valid solution, and in fact, many specific power series that arise from solving particular differential equations have been extensively studied and hold prominent places in mathematics and physics.

Introduction to Power Series

A **power series in x about the point x_0** is an expression of the form

$$c_0 + c_1(x - x_0) + c_2(x - x_0)^2 + \cdots$$

where the coefficients c_n are constants. This is concisely written using summation notation as follows:

$$\sum_{n=0}^{\infty} c_n(x - x_0)^n$$

Attention will be restricted to $x_0 = 0$; such series are simply called **power series in x:**

$$c_0 + c_1x + c_2x^2 + \cdots = \sum_{n=0}^{\infty} c_nx^n$$

A series is useful only if it **converges** (that is, if it approaches a finite limiting sum), so the natural question is, for what values of x will a given power series converge? Every power series in x falls into one of three categories:

Category 1: The power series converges only for $x = 0$.

Category 2: The power series converges for $|x| < R$ and **diverges** (that is, fails to converge) for $|x| > R$ (where R is some positive number).

Category 3: The power series converges for all x.

Since power series that converge only for $x = 0$ are essentially useless, only those power series that fall into Category 2 or Category 3 will be discussed here.

The **ratio test** says that the power series

$$\sum_{n=0}^{\infty} c_n x^n$$

will converge if

$$\lim_{n \to \infty} \left| \frac{c_{n+1} x^{n+1}}{c_n x^n} \right| < 1 \qquad (*)$$

and diverge if this limit is greater than 1. But (*) is equivalent to

$$|x| < \lim_{n \to \infty} \left| \frac{c_n}{c_{n+1}} \right|$$

so the positive number R mentioned in the definition of a Category 2 power series is this limit:

$$R = \lim_{n \to \infty} \left| \frac{c_n}{c_{n+1}} \right|$$

If this limit is ∞, then the power series converges for $|x| < \infty$—which means for all x—and the power series belongs to Category 3. R is called the **radius of convergence** of the power series, and the set of all x for which a real power series converges is always an interval, called its **interval of convergence.**

Example 1: Find the radius and interval of convergence for each of these power series:

$$\text{(a)} \quad \sum_{n=0}^{\infty} \frac{2^n}{n!} x^n \qquad \text{(b)} \quad \sum_{n=1}^{\infty} \frac{n^3}{3^n} x^n \qquad \text{(c)} \quad \sum_{n=1}^{\infty} \frac{1}{n} x^n$$

[Recall that $n!$ ("n factorial") denotes the product of the positive integers from 1 to n. For example, $4! = 1 \cdot 2 \cdot 3 \cdot 4 = 24$. By definition, $0!$ is set equal to 1.]

(a) In this power series, $c_n = 2^n/n!$, so the ratio test says

$$R = \lim_{n\to\infty} \left| \frac{c_n}{c_{n+1}} \right| = \lim_{n\to\infty} \left| \frac{2^n}{n!} \cdot \frac{(n+1)!}{2^{n+1}} \right| = \lim_{n\to\infty} \frac{n+1}{2} = \infty$$

Therefore, this series converges for all x.

(b) The radius of convergence of the power series in (b) is

$$R = \lim_{n\to\infty} \left| \frac{c_n}{c_{n+1}} \right| = \lim_{n\to\infty} \left| \frac{n^3}{3^n} \cdot \frac{3^{n+1}}{(n+1)^3} \right|$$

$$= \lim_{n\to\infty} \left[\frac{n^3}{(n+1)^3} \cdot \frac{3^{n+1}}{3^n} \right] = 1 \cdot 3 = 3$$

Since $R = 3$, the power series converges for $|x| < 3$ and diverges for $|x| > 3$. For a power series with a finite interval of convergence, the question of convergence at the endpoints of the interval must be examined separately. It may happen that the power series converges at neither endpoint, at only one, or at both. The power series

$$\sum_{n=1}^{\infty} \frac{n^3}{3^n} x^n$$

converges at neither the endpoint $x = 3$ nor $x = -3$ because the individual terms of both resulting series

$$\sum_{n=1}^{\infty} n^3 \quad \text{and} \quad \sum_{n=1}^{\infty} (-1)^n n^3$$

clearly do not approach 0 as $n \to \infty$. (For any series to converge, it is necessary that the individual terms go to 0.) Therefore, the interval of convergence of the power series in (b) is the open interval $-3 < x < 3$.

(c) The radius of convergence of this power series is

$$R = \lim_{n\to\infty} \left| \frac{c_n}{c_{n+1}} \right| = \lim_{n\to\infty} \left| \frac{1}{n} \cdot \frac{n+1}{1} \right| = \lim_{n\to\infty} \frac{n+1}{n} = 1$$

Since $R = 1$, the series

$$\sum_{n=1}^{\infty} \frac{1}{n} x^n$$

converges for $|x| < 1$ and diverges for $|x| > 1$. Since this power series has a finite interval of convergence, the question of convergence at the endpoints of the interval must be examined separately. At the endpoint $x = -1$, the power series becomes

$$\sum_{n=1}^{\infty} (-1)^n \frac{1}{n}$$

which converges, since it is an **alternating series** whose terms go to 0. However, at the endpoint $x = 1$, the power series becomes

$$\sum_{n=1}^{\infty} \frac{1}{n}$$

which is known to diverge (it is the **harmonic series**). Therefore, the interval of convergence of the power series

$$\sum_{n=1}^{\infty} \frac{1}{n} x^n$$

is the half-open interval $-1 \le x < 1$. ■

Taylor Series

A Category 2 or Category 3 power series in x defines a function f by setting

$$f(x) = c_0 + c_1x + c_2x^2 + c_3x^3 + c_4x^4 + \cdots = \sum_{n=0}^{\infty} c_nx^n \qquad (0)$$

for any x in the series' interval of convergence.

The power series expansion for $f(x)$ can be differentiated term by term, and the resulting series is a valid representation of $f'(x)$ in the same interval:

$$f'(x) = c_1 + 2c_2x + 3c_3x^2 + 4c_4x^3 + \cdots = \sum_{n=1}^{\infty} nc_nx^{n-1} \qquad (1)$$

Differentiating again gives

$$f''(x) = 2c_2 + 6c_3x + 12c_4x^2 + \cdots = \sum_{n=2}^{\infty} n(n-1)c_nx^{n-2} \qquad (2)$$

and so on. Substituting

$x = 0$ in equation (0) yields $c_0 = f(0)$,

$x = 0$ in equation (1) yields $c_1 = f'(0)$,

$x = 0$ in equation (2) yields $c_2 = \dfrac{f''(0)}{2}$,

and in general, substituting $x = 0$ in the power series expansion for the nth derivative of f yields

$$c_n = \frac{f^{(n)}(0)}{n!}$$

These are called the **Taylor coefficients** of f, and the resulting power series

$$\sum_{n=0}^{\infty} \frac{f^{(n)}(0)}{n!} x^n$$

is called the **Taylor series** of the function f.

Given a function f, its Taylor coefficients can be computed by the simple formula above, and the question arises, does the Taylor series of f actually converge to $f(x)$? If it does, that is, if

$$f(x) = \sum_{n=0}^{\infty} \frac{f^{(n)}(0)}{n!} x^n$$

for all x in some **neighborhood of** (interval around) 0, then the function f is said to be **analytic** (at 0). [More generally, if you form the Taylor series of f about a point $x = x_0$,

$$\sum_{n=0}^{\infty} \frac{f^{(n)}(x_0)}{n!} (x - x_0)^n$$

and if this series actually converges to $f(x)$ for all x in some neighborhood of x_0, then f is said to be analytic at x_0.] Polynomials are analytic everywhere, and rational functions (quotients of polynomials) are analytic at all points where the denominator is not zero. Furthermore, the familiar **transcendental** (that is, nonalgebraic) functions e^x, $\sin x$, and $\cos x$ are also analytic everywhere. The Taylor series in Table 4 are encountered so frequently that they are worth memorizing.

For a general power series (like the ones in Example 1), it is usually not possible to express it in closed form in terms of familiar functions.

Table 4
TAYLOR SERIES

$$f(x) \leftrightarrow \sum_{n=0}^{\infty} \frac{f^{(n)}(0)}{n!} x^n$$

$\dfrac{1}{1-x} = 1 + x + x^2 + x^3 + \cdots = \sum_{n=0}^{\infty} x^n$	(for $\lvert x\rvert < 1$)
$e^x = 1 + x + \dfrac{x^2}{2!} + \dfrac{x^3}{3!} + \cdots = \sum_{n=0}^{\infty} \dfrac{x^n}{n!}$	(for all x)
$\sin x = x - \dfrac{x^3}{3!} + \dfrac{x^5}{5!} - \cdots = \sum_{n=0}^{\infty} (-1)^n \dfrac{x^{2n+1}}{(2n+1)!}$	(for all x)
$\cos x = 1 - \dfrac{x^2}{2!} + \dfrac{x^4}{4!} - \cdots = \sum_{n=0}^{\infty} (-1)^n \dfrac{x^{2n}}{(2n)!}$	(for all x)

Example 2: Use Table 4 to find the Taylor series expansion of each of the following functions:

(a) $\dfrac{1}{1-x^2}$ (b) $\dfrac{1}{(1-x)^2}$ (c) $\ln(1+x)$

(d) e^{-x^2} (e) $x \cos x$ (f) $\sin x \cos x$

(g) $\arctan x$

(a) Replacing x by x^2 in the Taylor series expansion of $1/(1-x)$ gives

$$\frac{1}{1-x^2} = 1 + x^2 + (x^2)^2 + (x^2)^3 + \cdots \quad \text{for } |x^2| < 1$$

$$= 1 + x^2 + x^4 + x^6 + \cdots$$

$$= \sum_{n=0}^{\infty} x^{2n} \quad \text{for } |x| < 1$$

since $|x| < 1$ is equivalent to $|x^2| < 1$.

(b) Differentiating $1/(1-x)$ gives $1/(1-x)^2$, so differentiating the Taylor series expansion of $1/(1-x)$ term by term will give the series expansion of $1/(1-x)^2$:

$$\frac{1}{(1-x)^2} = \frac{d}{dx}\left(\frac{1}{1-x}\right) = \frac{d}{dx}\sum_{n=0}^{\infty} x^n = \sum_{n=0}^{\infty} \frac{d}{dx}(x^n) = \sum_{n=1}^{\infty} nx^{n-1}$$

$$\text{for } |x| < 1$$

(c) First, replacing x by $-x$ in the Taylor series expansion of $1/(1-x)$ gives the expansion of $1/(1+x)$:

$$\frac{1}{1+x} = \frac{1}{1-(-x)} = 1 + (-x) + (-x)^2 + (-x)^3 + \cdots \quad \text{for } |-x| < 1$$

$$= \sum_{n=0}^{\infty} (-1)^n x^n \quad \text{for } |x| < 1$$

Now, since integrating $1/(1+x)$ yields $\ln(1+x)$, integrating the Taylor series for $1/(1+x)$ term by term gives the expansion for $\ln(1+x)$, valid for $|x| < 1$:

$$\ln(1+x) = \int \frac{1}{1+x}\,dx = \int \sum_{n=0}^{\infty} (-1)^n x^n\,dx = \sum_{n=0}^{\infty} \int (-1)^n x^n\,dx$$

$$= \sum_{n=0}^{\infty} (-1)^n \frac{x^{n+1}}{n+1}$$

Technical note: Integrating $1/(1+x)$ yields $\ln(1+x) + c$ (where c is some arbitrary constant), so strictly speaking, the equation above should have been written

$$\ln(1+x) + c = \sum_{n=0}^{\infty} (-1)^n \frac{x^{n+1}}{n+1}$$

However, substituting $x = 0$ into this equation shows that $c = 0$, so the expansion given above for $\ln(1+x)$ is indeed correct.

(d) Replacing x by $-x^2$ in the Taylor series expansion of e^x yields the desired result:

$$e^{-x^2} = \sum_{n=0}^{\infty} \frac{(-x^2)^n}{n!} = \sum_{n=0}^{\infty} (-1)^n \frac{x^{2n}}{n!}$$

(e) Multiplying each term of the Taylor series for $\cos x$ by x gives

$$x \cos x = \sum_{n=0}^{\infty} \frac{(-1)^n}{(2n)!} x^{2n+1}$$

(f) One way to find the series expansion for $\sin x \cos x$ is to multiply the expansions of $\sin x$ and $\cos x$. A faster way, however, involves recalling the trigonometric identity $\sin 2x = 2 \sin x \cos x$ and then replacing x by $2x$ in the series expansion of $\sin x$:

$$\sin x \cos x = \frac{1}{2} \sin 2x$$

$$= \frac{1}{2} \sum_{n=0}^{\infty} (-1)^n \frac{(2x)^{2n+1}}{(2n+1)!}$$

$$= \sum_{n=0}^{\infty} \frac{(-1)^n 2^{2n}}{(2n+1)!} x^{2n+1}$$

(g) Since arctan x is the integral of $1/(1 + x^2)$, integrate the series expansion of $1/(1 + x^2)$ term by term:

$$\arctan x = \int \frac{1}{1 + x^2}\, dx$$

$$= \int \frac{1}{1 - (-x^2)}\, dx$$

$$= \int \left[\sum_{n=0}^{\infty} (-x^2)^n \right] dx \quad \text{for } |x^2| < 1$$

$$= \sum_{n=0}^{\infty} \left[\int (-1)^n x^{2n}\, dx \right]$$

$$= \sum_{n=0}^{\infty} (-1)^n \frac{x^{2n+1}}{2n+1} \quad \text{for } |x| < 1$$

Recall the technical note accompanying part (c) above (which also involved the term-by-term integration of a power series). The integral of $1/(1 + x^2)$ is actually arctan $x + c$, and the equation above should read

$$\arctan x + c = \sum_{n=0}^{\infty} (-1)^n \frac{x^{2n+1}}{2n+1}$$

However, substituting $x = 0$ into this equation shows that $c = 0$, so the expansion given above for arctan x is indeed correct. ■

Power Series Solutions of Differential Equations

First-order equations. The validity of term-by-term differentiation of a power series within its interval of convergence implies that first-order differential equations may be solved by assuming a solution of the form

$$y = \sum_{n=0}^{\infty} c_n x^n$$

substituting this into the equation, and then determining the coefficients c_n.

Example 3: Find a power series solution of the form

$$y = \sum_{n=0}^{\infty} c_n x^n$$

for the differential equation

$$y' - xy = 0$$

Substituting

$$y = \sum_{n=0}^{\infty} c_n x^n$$

into the differential equation yields

$$\sum_{n=1}^{\infty} nc_n x^{n-1} - \sum_{n=0}^{\infty} c_n x^{n+1} = 0$$

Now, write out the first few terms of each series,

$$(c_1 + 2c_2x + 3c_3x^2 + 4c_4x^3 + \cdots) - (c_0x + c_1x^2 + c_2x^3 + \cdots) = 0$$

and combine like terms:

$$c_1 + (2c_2 - c_0)x + (3c_3 - c_1)x^2 + (4c_4 - c_2)x^3 + \cdots = 0$$

Since the pattern is clear, this last equation may be written as

$$c_1 + \sum_{n=2}^{\infty} (nc_n - c_{n-2})x^{n-1} = 0$$

In order for this equation to hold true for all x, every coefficient on the left-hand side must be zero. This means $c_1 = 0$, and for all $n \geq 2$,

$$nc_n - c_{n-2} = 0$$

This last equation defines the **recurrence relation** that holds for the coefficients of the power series solution:

$$c_n = \frac{c_{n-2}}{n} \quad \text{for } n \geq 2$$

Since there is no constraint on c_0, c_0 is an arbitrary constant, and it is already known that $c_1 = 0$. The recurrence relation above says $c_2 = \frac{1}{2}c_0$ and $c_3 = \frac{1}{3}c_1$, which equals 0 (because c_1 does). In fact, it is easy to see that every coefficient c_n with n odd will be zero. As for c_4, the recurrence relation says

$$c_4 = \frac{c_2}{4} = \frac{c_0}{2 \cdot 4}$$

and so on. Since all c_n with n odd equal 0, the desired power series solution is therefore

$$\begin{aligned} y &= c_0 + c_2x^2 + c_4x^4 + c_6x^6 + \cdots \\ &= c_0 + \frac{c_0}{2}x^2 + \frac{c_0}{2 \cdot 4}x^4 + \frac{c_0}{2 \cdot 4 \cdot 6}x^6 + \cdots \\ &= c_0\left(1 + \frac{1}{2}x^2 + \frac{1}{2 \cdot 4}x^4 + \frac{1}{2 \cdot 4 \cdot 6}x^6 + \cdots\right) \\ &= c_0 \sum_{n=0}^{\infty} \frac{1}{n!2^n}x^{2n} \end{aligned}$$

Note that the general solution contains one parameter (c_0), as expected for a first-order differential equation. This power series is unusual in that it is possible to express it in terms of an elementary function. Observe:

$$c_0 \sum_{n=0}^{\infty} \frac{1}{n!2^n} x^{2n} = c_0 \sum_{n=0}^{\infty} \frac{1}{n!}\left(\frac{1}{2}x^2\right)^n = c_0 e^{x^2/2}$$

It is easy to check that $y = c_0 e^{x^2/2}$ is indeed the solution of the given differential equation, $y' = xy$. Remember: Most power series cannot be expressed in terms of familiar, elementary functions, so the final answer would be left in the form of a power series. ■

Example 4: Find a power series expansion for the solution of the IVP

$$y' = x + y$$

$$y(0) = 1$$

Substituting

$$y = \sum_{n=0}^{\infty} c_n x^n$$

into the differential equation yields

$$\sum_{n=1}^{\infty} nc_n x^{n-1} = x + \sum_{n=0}^{\infty} c_n x^n$$

or, collecting all the terms on one side,

$$\sum_{n=1}^{\infty} nc_n x^{n-1} - x - \sum_{n=0}^{\infty} c_n x^n = 0$$

Writing out the first few terms of the series yields

$$(c_1 + 2c_2x + 3c_3x^2 + 4c_4x^3 + \cdots) - x - (c_0 + c_1x + c_2x^2 + c_3x^3 + \cdots) = 0$$

or, upon combining like terms,

$$(c_1 - c_0) + (2c_2 - 1 - c_1)x + (3c_3 - c_2)x^2 + (4c_4 - c_3)x^3 + \cdots = 0$$

Now that the pattern is clear, this last equation can be written

$$(c_1 - c_0) + (2c_2 - 1 - c_1)x + \sum_{n=3}^{\infty} (nc_n - c_{n-1})x^{n-1} = 0$$

In order for this equation to hold true for all x, every coefficient on the left-hand side must be zero. This means

$$c_1 - c_0 = 0, \quad 2c_2 - 1 - c_1 = 0, \quad \text{and} \quad nc_n - c_{n-1} = 0 \quad \text{for } n \geq 3 \qquad (*)$$

The last equation defines the recurrence relation that determines the coefficients of the power series solution:

$$c_n = \frac{c_{n-1}}{n} \quad \text{for } n \geq 3$$

The first equation in (*) says $c_1 = c_0$, and the second equation says $c_2 = \frac{1}{2}(1 + c_1) = \frac{1}{2}(1 + c_0)$. Next, the recurrence relation says

$$c_3 = \frac{c_2}{3} = \frac{\frac{1}{2}(1 + c_0)}{3} = \frac{1 + c_0}{2 \cdot 3}$$

$$c_4 = \frac{c_3}{4} = \frac{1 + c_0}{2 \cdot 3 \cdot 4}$$

and so on. Collecting all these results, the desired power series solution is therefore

$$y = c_0 + c_1x + c_2x^2 + c_3x^3 + c_4x^4 + \cdots$$

$$= c_0 + c_0x + \frac{1 + c_0}{2}x^2 + \frac{1 + c_0}{2 \cdot 3}x^3 + \frac{1 + c_0}{2 \cdot 3 \cdot 4}x^4 + \cdots$$

$$= c_0 + c_0x + (1 + c_0)\left(\frac{1}{2}x^2 + \frac{1}{2 \cdot 3}x^3 + \frac{1}{2 \cdot 3 \cdot 4}x^4 + \cdots\right)$$

$$= c_0 + c_0x + (1 + c_0)\sum_{n=2}^{\infty}\frac{1}{n!}x^n$$

Now, the initial condition is applied to evaluate the parameter c_0:

$$y(0) = 1 \Rightarrow \left[c_0 + c_0x + (1 + c_0)\sum_{n=2}^{\infty}\frac{1}{n!}x^n\right]_{x=0} = 1 \Rightarrow c_0 = 1$$

Therefore, the power series expansion for the solution of the given IVP is

$$y = 1 + x + \sum_{n=2}^{\infty}\frac{2}{n!}x^n \qquad (**)$$

If desired, it is possible to express this in terms of elementary functions. Since

$$\sum_{n=0}^{\infty}\frac{2}{n!}x^n = \frac{2}{0!}x^0 + \frac{2}{1!}x^1 + \sum_{n=2}^{\infty}\frac{2}{n!}x^n$$

equation (**) may be written

$$\begin{aligned} y &= 1 + x + \sum_{n=2}^{\infty}\frac{2}{n!}x^n \\ &= 1 + x + \left[\sum_{n=0}^{\infty}\frac{2}{n!}x^n - \frac{2}{0!}x^0 - \frac{2}{1!}x^1\right] \\ &= 1 + x + (2e^x - 2 - 2x) \\ &= 2e^x - x - 1 \end{aligned}$$

which does indeed satisfy the given IVP, as you can readily verify. ■

Second-order equations. The process of finding power series solutions of homogeneous second-order linear differential equations is more subtle than for first-order equations. Any homogeneous second-order linear differential equation may be written in the form

$$y'' + p(x)y' + q(x)y = 0$$

If both coefficient functions p and q are analytic at x_0, then x_0 is called an **ordinary point** of the differential equation. On the other hand, if even one of these functions fails to be analytic at x_0, then x_0 is called a **singular point.** Since the method for finding a solution that is a power series in x_0 is considerably more complicated if x_0 is a singular point, attention here will be restricted to power series solutions at ordinary points.

Example 5: Find a power series solution in x for the IVP

$$\begin{aligned} y'' - xy' + y &= 0 \\ y(0) &= 2 \\ y'(0) &= 3 \end{aligned}$$

Substituting

$$y = \sum_{n=0}^{\infty} c_n x^n$$

into the differential equation yields

$$\sum_{n=2}^{\infty} n(n-1)c_n x^{n-2} - \sum_{n=1}^{\infty} nc_n x^n + \sum_{n=0}^{\infty} c_n x^n = 0 \qquad (*)$$

The solution may now proceed as in the examples above, writing out the first few terms of the series, collecting like terms, and then determining the constraints on the coefficients from the emerging pattern. Here's another method.

The first step is to re-index the series so that each one involves x^n. In the present case, only the first series must be subjected to this procedure. Replacing n by $n + 2$ in this series yields

$$\sum_{n=2}^{\infty} n(n-1)c_n x^{n-2} = \sum_{n+2=2}^{\infty} (n+2)[(n+2)-1]c_{n+2}x^{(n+2)-2}$$

$$= \sum_{n=0}^{\infty} (n+2)(n+1)c_{n+2}x^n$$

Therefore, equation (*) becomes

$$\sum_{n=0}^{\infty} (n+2)(n+1)c_{n+2}x^n - \sum_{n=1}^{\infty} nc_n x^n + \sum_{n=0}^{\infty} c_n x^n = 0 \qquad (**)$$

The next step is to rewrite the left-hand side in terms of a *single* summation. The index n ranges from 0 to ∞ in the first and third series, but only from 1 to ∞ in the second. Since the common range of all the series is therefore 1 to ∞, the single summation which will help replace the left-hand side will range from 1 to ∞. Consequently, it is necessary to first write (**) as

$$\left[2c_2 + \sum_{n=1}^{\infty} (n+2)(n+1)c_{n+2}x^n\right] - \sum_{n=1}^{\infty} nc_n x^n + \left[c_0 + \sum_{n=1}^{\infty} c_n x^n\right] = 0$$

and then combine the series into a single summation:

$$(2c_2 + c_0) + \sum_{n=1}^{\infty} [(n+2)(n+1)c_{n+2} - (n-1)c_n]x^n = 0$$

In order for this equation to hold true for all x, every coefficient on the left-hand side must be zero. This means $2c_2 + c_0 = 0$, and for $n \geq 1$, the following recurrence relation holds:

$$c_{n+2} = \frac{n-1}{(n+1)(n+2)}c_n$$

Since there is no restriction on c_0 or c_1, these will be arbitrary, and the equation $2c_2 + c_0 = 0$ implies $c_2 = -\frac{1}{2}c_0$. For the coefficients from c_3 on, the recurrence relation is needed:

$$c_3 = c_{1+2} = \frac{1-1}{(1+1)(1+2)}c_1 = 0$$

$$c_4 = c_{2+2} = \frac{2-1}{(2+1)(2+2)}c_2 = \frac{1}{3\cdot 4}\cdot\frac{-c_0}{2}$$

$$c_5 = c_{3+2} = \frac{3-1}{(3+1)(3+2)}c_3 = \frac{2}{4\cdot 5}\cdot 0 = 0$$

$$c_6 = c_{4+2} = \frac{4-1}{(4+1)(4+2)}c_4 = \frac{3}{5\cdot 6}\cdot\frac{-c_0}{2\cdot 3\cdot 4}$$

$$c_7 = c_{5+2} = \frac{5-1}{(5+1)(5+2)}c_5 = \frac{4}{6\cdot 7}\cdot 0 = 0$$

$$\vdots$$

The pattern here isn't too difficult to discern: $c_n = 0$ for all odd $n \geq 3$, and for all even $n \geq 4$,

$$c_n = -\frac{n-3}{n!}c_0$$

This recurrence relation can be restated as follows: for all $n \geq 2$,

$$c_{2n} = -\frac{2n-3}{(2n)!}c_0 = \frac{3-2n}{(2n)!}c_0$$

The desired power series solution is therefore

$$\begin{aligned} y &= c_0 + c_1x + c_2x^2 + c_3x^3 + c_4x^4 + c_5x^5 + c_6x^6 + \cdots \\ &= c_0 + c_1x - \frac{c_0}{2}x^2 + 0 - \frac{c_0}{4!}x^4 + 0 - \frac{3c_0}{6!}x^6 + \cdots \end{aligned}$$

$$= c_0\left[1 - \frac{1}{2}x^2 - \frac{1}{4!}x^4 - \frac{3}{6!}x^6 - \cdots\right] + c_1x$$

$$= c_0\left[1 - \frac{1}{2}x^2 + \sum_{n=2}^{\infty} \frac{3-2n}{(2n)!}x^{2n}\right] + c_1x$$

As expected for a second-order differential equation, the general solution contains two parameters (c_0 and c_1), which will be determined by the initial conditions. Since $y(0) = 2$, it is clear that $c_0 = 2$, and then, since $y'(0) = 3$, the value of c_1 must be 3. The solution of the given IVP is therefore

$$y = 2 + 3x - x^2 + \sum_{n=2}^{\infty} \frac{2(3-2n)}{(2n)!}x^{2n} \quad \blacksquare$$

Example 6: Find a power series solution in x for the differential equation

$$(x^2 + 1)y'' + y' - x^2y = 0$$

Substituting

$$y = \sum_{n=0}^{\infty} c_nx^n$$

into the given equation yields

$$(x^2 + 1)\sum_{n=2}^{\infty} n(n-1)c_nx^{n-2} + \sum_{n=1}^{\infty} nc_nx^{n-1} - \sum_{n=0}^{\infty} c_nx^{n+2} = 0$$

or

$$\sum_{n=2}^{\infty} n(n-1)c_nx^n + \sum_{n=2}^{\infty} n(n-1)c_nx^{n-2} + \sum_{n=1}^{\infty} nc_nx^{n-1}$$

$$- \sum_{n=0}^{\infty} c_nx^{n+2} = 0 \qquad (*)$$

Now, all series but the first must be re-indexed so that each involves x^n:

$$\sum_{n=2}^{\infty} n(n-1)c_n x^{n-2} = \sum_{n+2=2}^{\infty} (n+2)[(n+2)-1]c_{n+2}x^{(n+2)-2}$$

$$= \sum_{n=0}^{\infty} (n+2)(n+1)c_{n+2}x^n$$

$$\sum_{n=1}^{\infty} nc_n x^{n-1} = \sum_{n+1=1}^{\infty} (n+1)c_{n+1}x^{(n+1)-1} = \sum_{n=0}^{\infty} (n+1)c_{n+1}x^n$$

$$\sum_{n=0}^{\infty} c_n x^{n+2} = \sum_{n-2=0}^{\infty} c_{n-2}x^{(n-2)+2} = \sum_{n=2}^{\infty} c_{n-2}x^n$$

Therefore, equation (*) becomes

$$\sum_{n=2}^{\infty} n(n-1)c_n x^n + \sum_{n=0}^{\infty} (n+2)(n+1)c_{n+2}x^n + \sum_{n=0}^{\infty} (n+1)c_{n+1}x^n - \sum_{n=2}^{\infty} c_{n-2}x^n = 0 \qquad (**)$$

The next step is to rewrite the left-hand side in terms of a *single* summation. The index n ranges from 0 to ∞ in the second and third series, but only from 2 to ∞ in the first and fourth. Since the common range of all the series is therefore 2 to ∞, the single summation which will help replace the left-hand side will range from 2 to ∞. It is therefore necessary to first write (**) as

$$\sum_{n=2}^{\infty} n(n-1)c_n x^n + \left[2c_2 + 6c_3 x + \sum_{n=2}^{\infty} (n+2)(n+1)c_{n+2}x^n\right] + \left[c_1 + 2c_2 x + \sum_{n=2}^{\infty} (n+1)c_{n+1}x^n\right] - \sum_{n=2}^{\infty} c_{n-2}x^n = 0$$

and then combine the series into a single summation:

$$(c_1 + 2c_2) + (2c_2 + 6c_3)x$$

$$+ \sum_{n=2}^{\infty} [n(n-1)c_n + (n+2)(n+1)c_{n+2} + (n+1)c_{n+1} - c_{n-2}]x^n = 0$$

Again, in order for this equation to hold true for all x, every coefficient on the left-hand side must be zero. This means $c_1 + 2c_2 = 0$, $2c_2 + 6c_3 = 0$, and for $n \geq 2$, the following recurrence relation holds:

$$c_{n+2} = \frac{c_{n-2} - n(n-1)c_n - (n+1)c_{n+1}}{(n+1)(n+2)}$$

Since there is no restriction on c_0 or c_1, these will be arbitrary; the equation $c_1 + 2c_2 = 0$ implies $c_2 = -\frac{1}{2}c_1$, and the equation $2c_2 + 6c_3 = 0$ implies $c_3 = -\frac{1}{3}c_2 = -\frac{1}{3}(-\frac{1}{2}c_1) = \frac{1}{6}c_1$. For the coefficients from c_4 on, the recurrence relation is needed:

$$c_4 = c_{2+2} = \frac{c_{2-2} - 2(2-1)c_2 - (2+1)c_{2+1}}{(2+1)(2+2)}$$

$$= \frac{c_0 - 2c_2 - 3c_3}{3 \cdot 4} = \frac{c_0 - 2(-\frac{1}{2}c_1) - 3(\frac{1}{6}c_1)}{3 \cdot 4} = \frac{1}{12}c_0 + \frac{1}{24}c_1$$

$$c_5 = c_{3+2} = \frac{c_{3-2} - 3(3-1)c_3 - (3+1)c_{3+1}}{(3+1)(3+2)}$$

$$= \frac{c_1 - 6c_3 - 4c_4}{4 \cdot 5} = \frac{c_1 - 6(\frac{1}{6}c_1) - 4(\frac{1}{12}c_0 + \frac{1}{24}c_1)}{4 \cdot 5}$$

$$= -\frac{1}{60}c_0 - \frac{1}{120}c_1$$

$$c_6 = c_{4+2} = \frac{c_{4-2} - 4(4-1)c_4 - (4+1)c_{4+1}}{(4+1)(4+2)}$$

$$= \frac{c_2 - 12c_4 - 5c_5}{5 \cdot 6}$$

$$= \frac{-\frac{1}{2}c_1 - 12(\frac{1}{12}c_0 + \frac{1}{24}c_1) - 5(-\frac{1}{60}c_0 - \frac{1}{120}c_1)}{5 \cdot 6}$$

$$= -\tfrac{11}{360}c_0 - \tfrac{23}{720}c_1$$

$$\vdots$$

The desired power series solution is therefore

$$y = c_0 + c_1x + c_2x^2 + c_3x^3 + c_4x^4 + c_5x^5 + c_6x^6 + \cdots$$

$$= c_0 + c_1x + (-\tfrac{1}{2}c_1)x^2 + (\tfrac{1}{6}c_1)x^3 + (\tfrac{1}{12}c_0 + \tfrac{1}{24}c_1)x^4$$

$$+ (-\tfrac{1}{60}c_0 - \tfrac{1}{120}c_1)x^5 + (-\tfrac{11}{360}c_0 - \tfrac{23}{720}c_1)x^6 + \cdots$$

$$= c_0(1 + \tfrac{1}{12}x^4 - \tfrac{1}{60}x^5 - \tfrac{11}{360}x^6 - \cdots)$$

$$+ c_1(x - \tfrac{1}{2}x^2 + \tfrac{1}{6}x^3 + \tfrac{1}{24}x^4 - \tfrac{1}{120}x^5 - \tfrac{23}{720}x^6 - \cdots)$$

Determining a specific pattern to these coefficients would be a tedious exercise (note how complicated the recurrence relation is), so the final answer is simply left in this form. ■

If you are faced with an IVP that involves a linear differential equation with constant coefficients, you can proceed by the method of undetermined coefficients or variation of parameters and then apply the initial conditions to evaluate the constants. However, what if the nonhomogeneous right-hand term is discontinuous? There exists a method for solving such problems that can also be used to solve less frightening IVP's (that is, ones that do not involve discontinuous terms) and even some equations whose coefficients are not constants. One of the features of this alternative method for solving IVP's is that the values of the parameters are not found after the general solution has been obtained. Instead, the initial conditions are incorporated right into the initial stages of the solution, so when the final step is completed, the arbitrary constants have already been evaluated.

Linear Transformations

A function is usually introduced as a rule which acts on a number to produce a unique numerical result. That is, a function accepts a number as input and produces a number as output. For instance, consider the function defined by the equation $f(x) = x^2$; it specifies a particular operation to be performed on any given value of x. When this function acts on the number 3, for example, it gives the result 9, a fact which can be symbolized as follows:

$$3 \overset{f}{\mapsto} 9$$

However, functions are not restricted to acting only on numbers to produce other numbers. A function can also act on a *function* to produce *another function*. These "superfunctions" are often referred to as **operators** or **transformations.** Therefore, an operator accepts a *function* as input and produces a *function* as output.

A familiar example is the **differentiation operator,** D:

$$f(x) \overset{D}{\mapsto} f'(x)$$

This operator acts on a (differentiable) function to produce another function: namely, the derivative of the input function. For example,

$$x^3 \overset{D}{\mapsto} 3x^2, \qquad \sin x \overset{D}{\mapsto} \cos x, \qquad e^x \overset{D}{\mapsto} e^x$$

Another well-known example is the **integration operator,** which acts on an (integrable) function to produce another function: its integral. Since an operator is a function, it must produce one and only one output for each input; therefore, it makes sense here to consider an integration operator I of the form

$$f \overset{I}{\mapsto} \int_0^x f(t)\,dt$$

[The letter t, a dummy variable, is chosen simply to distinguish it from x, the upper limit of integration. It's usually considered bad form to write, for example, $\int_0^x f(x)\,dx$.]

To illustrate, if $f(x) = x^2$, then

$$I[f] = \int_0^x f(t)\,dt = \int_0^x t^2\,dt = \tfrac{1}{3}x^3$$

that is,

$$x^2 \overset{I}{\mapsto} \tfrac{1}{3}x^3$$

Both of these operators, differentiation D and integration I, enjoy an important property known as **linearity.** Any operator or transformation T is said to be **linear** if both of the following conditions,

$$T[c \cdot f] = c \cdot T[f]$$

$$T[f + g] = T[f] + T[g]$$

always hold for all constants c and all admissible functions f and g. Since

> the derivative of a constant multiple of a function is equal to the constant times the derivative of the function,

and since

> the derivative of the sum of two functions is the sum of the derivatives,

differentiation is indeed linear. Furthermore, since both of the previous statements remain true when "derivative" is replaced by "integral," integration is also a linear transformation.

The Laplace Transform Operator

A particular kind of integral transformation is known as the **Laplace transformation,** denoted by L. The definition of this operator is

$$L[f(x)] = \int_0^\infty e^{-px} f(x)\, dx$$

The result—called the **Laplace transform** of f—will be a function of p, so in general,

$$f(x) \stackrel{L}{\mapsto} F(p)$$

Example 1: Find the Laplace transform of the function $f(x) = x$.

By definition,

$$L[x] = \int_0^\infty e^{-px}\,x\,dx$$

Integrating by parts yields

$$\begin{aligned}
L[x] &= \int_0^\infty e^{-px}\,x\,dx \\
&= -\frac{x}{p}e^{-px}\Big]_0^\infty - \int_0^\infty -\frac{1}{p}e^{-px}\,dx \\
&= 0 + \frac{1}{p}\int_0^\infty e^{-px}\,dx \quad (\text{for } p > 0) \\
&= \frac{1}{p}\left[-\frac{1}{p}e^{-px}\right]_0^\infty \\
&= \frac{1}{p^2}
\end{aligned}$$

Therefore, the function $F(p) = 1/p^2$ is the Laplace transform of the function $f(x) = x$. [Technical note: The convergence of the improper integral here depends on p being positive, since only then will $(x/p)e^{-px}$ and e^{-px} approach a finite limit (namely 0) as $x \to \infty$. Therefore, the Laplace transform of $f(x) = x$ is defined only for $p > 0$.] ■

In general, it can be shown that for any nonnegative integer n,

$$\text{the Laplace transform of } \quad f(x) = x^n \quad \text{is} \quad F(p) = \frac{n!}{p^{n+1}}$$

Like the operators D and I—indeed, like all operators—the Laplace transform operator L acts on a function to produce another function. Furthermore, since

$$L[cf] = \int_0^\infty e^{-px} \cdot cf(x)\,dx = c\int_0^\infty e^{-px} f(x)\,dx = cL[f]$$

and

$$\begin{aligned} L[f+g] &= \int_0^\infty e^{-px}[f(x) + g(x)]\,dx \\ &= \int_0^\infty [e^{-px} f(x) + e^{-px}g(x)]\,dx \\ &= \int_0^\infty e^{-px} f(x)\,dx + \int_0^\infty e^{-px} g(x)\,dx \\ &= L[f] + L[g] \end{aligned}$$

the Laplace transform operator L is also linear.

[Technical note: Just as not all functions have derivatives or integrals, not all functions have Laplace transforms. For a function f to have a Laplace transform, it is sufficient that $f(x)$ be continuous (or at least piecewise continuous) for $x \geq 0$ and of **exponential order** (which means that for some constants c and λ, the inequality

$$|f(x)| \leq ce^{\lambda x}$$

holds for all x). Any **bounded** function (that is, any function f that always satisfies $|f(x)| \leq M$ for some $M \geq 0$) is automatically of exponential order (just take $c = M$ and $\lambda = 0$ in the defining inequality). Therefore, $\sin kx$ and $\cos kx$ each have a Laplace transform, since they are continuous and bounded functions. Furthermore, any function of the form e^{kx}, as well as any polynomial, is continuous and, although unbounded, is of exponential order and therefore has a Laplace transform. In short, most of the functions you are likely to encounter in practice will have Laplace transforms.]

Example 2: Find the Laplace transform of the function $f(x) = x^3 - 4x + 2$.

Recall from the first statement following Example 1 that the Laplace transform of $f(x) = x^n$ is $F(p) = n!/p^{n+1}$. Therefore, since the Laplace transform operator L is linear,

$$\begin{aligned} L[x^3 - 4x + 2] &= L[x^3] + L[-4x] + L[2] \\ &= L[x^3] - 4L[x^1] + 2L[x^0] \\ &= \frac{3!}{p^{3+1}} - 4\frac{1!}{p^{1+1}} + 2\frac{0!}{p^{0+1}} \\ F(p) &= \frac{6}{p^4} - \frac{4}{p^2} + \frac{2}{p} \quad \blacksquare \end{aligned}$$

Example 3: Determine the Laplace transform of $f(x) = e^{kx}$.

Apply the definition and perform the integration:

$$\begin{aligned} L[e^{kx}] &= \int_0^\infty e^{-px}e^{kx}\,dx \\ &= \int_0^\infty e^{-(p-k)x}\,dx \\ &= -\frac{1}{p-k}e^{-(p-k)x}\Big|_0^\infty \end{aligned}$$

In order for this improper integral to converge, the coefficient $(p - k)$ in the exponential must be positive (recall the technical note in Example 1). Thus, for $p > k$, the calculation yields

$$L[e^{kx}] = \frac{1}{p-k} \quad \blacksquare$$

Example 4: Find the Laplace transform of $f(x) = \sin kx$.

By definition,

$$L[\sin kx] = \int_0^\infty e^{-px} \sin kx \, dx$$

This integral is evaluated by performing integration by parts twice, as follows:

$$\begin{aligned}
\int e^{-px} \sin kx \, dx &= -\frac{1}{p} e^{-px} \sin kx - \int -\frac{1}{p} e^{-px} \cdot k \cos kx \, dx \\
&= -\frac{1}{p} e^{-px} \sin kx + \frac{k}{p} \int e^{-px} \cos kx \, dx \\
&= -\frac{1}{p} e^{-px} \sin kx + \\
&\quad \frac{k}{p} \left[-\frac{1}{p} e^{-px} \cos kx + \int \frac{1}{p} e^{-px} \cdot (-k \sin kx) \, dx \right] \\
&= -\frac{1}{p} e^{-px} \sin kx - \frac{k}{p^2} e^{-px} \cos kx \\
&\quad - \frac{k^2}{p^2} \int e^{-px} \sin kx \, dx
\end{aligned}$$

so

$$\left(1 + \frac{k^2}{p^2}\right) \int e^{-px} \sin kx \, dx = -\frac{1}{p} e^{-px} \sin kx - \frac{k}{p^2} e^{-px} \cos kx$$

$$\int e^{-px} \sin kx \, dx = \frac{-\frac{1}{p} e^{-px} \sin kx - \frac{k}{p^2} e^{-px} \cos kx}{1 + \frac{k^2}{p^2}}$$

$$\int e^{-px} \sin kx \, dx = \frac{-e^{-px}(p \sin kx + k \cos kx)}{p^2 + k^2}$$

Therefore,

$$L[\sin kx] = \int_0^\infty e^{-px}\sin kx\,dx = \frac{-e^{-px}(p\sin kx + k\cos kx)}{p^2+k^2}\bigg]_0^\infty = \frac{k}{p^2+k^2}$$

for $p > 0$. By a similar calculation, it can be shown that

$$L[\cos kx] = \frac{p}{p^2+k^2} \qquad \blacksquare$$

Example 5: Determine the Laplace transform of the function

$$f(x) = \begin{cases} 1 & \text{for } 0 \le x \le 2 \\ 0 & \text{for } x > 2 \end{cases}$$

pictured in Figure 8:

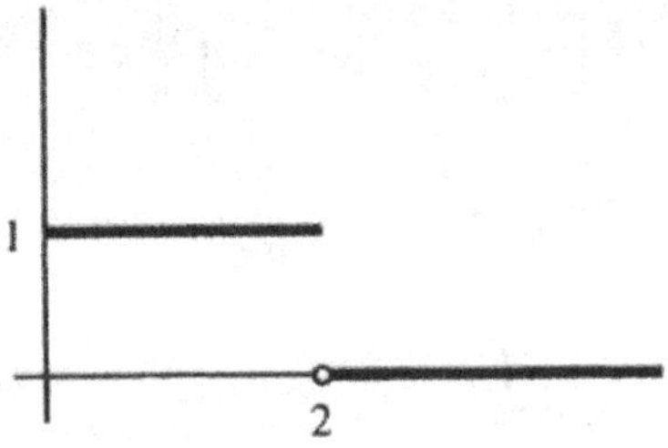

■ Figure 8 ■

This is an example of a **step function.** It is not continuous, but it is *piecewise* continuous, and since it is bounded, it is certainly of exponential order. Therefore, it has a Laplace transform.

$$\begin{aligned} L[f(x)] &= \int_0^\infty e^{-px} f(x)\,dx \\ &= \int_0^2 e^{-px} f(x)\,dx + \int_2^\infty e^{-px} f(x)\,dx \\ &= \int_0^2 e^{-px}\cdot 1\,dx + \int_2^\infty e^{-px}\cdot 0\,dx \\ &= \int_0^2 e^{-px}\,dx \\ &= -\frac{1}{p}e^{-px}\Big]_0^2 \\ L[f(x)] &= \frac{1-e^{-2p}}{p} \end{aligned}$$ ■

Table 5 (page 144) assembles the Laplace transforms of a few of the most frequently encountered functions, as well as some of the important properties of the Laplace transform operator L.

Example 6: Use Table 5 to find the Laplace transform of $f(x) = \sin^2 x$.

Invoking the trigonometric identity

$$\sin^2 x = \tfrac{1}{2}(1 - \cos 2x)$$

linearity of L implies

$$L[\sin^2 x] = \tfrac{1}{2}L[1 - \cos 2x] = \frac{1}{2}\left(\frac{1}{p} - \frac{p}{p^2+4}\right) = \frac{2}{p(p^2+4)}$$ ■

Table 5
LAPLACE TRANSFORMS

$f(x)$	$\xrightarrow{L}$	$F(p)$
1		$\dfrac{1}{p}$
x^n		$\dfrac{n!}{p^{n+1}}$
e^{kx}		$\dfrac{1}{p-k}$
$\sin kx$		$\dfrac{k}{p^2+k^2}$
$\cos kx$		$\dfrac{p}{p^2+k^2}$

Linearity:

$$L[c_1 f(x) + c_2 g(x)] = c_1 L[f(x)] + c_2 L[g(x)]$$

Shifting formula:

$$L[e^{kx} f(x)] = F(p-k), \quad \text{where } L[f(x)] = F(p)$$

Laplace transform of derivatives:

$$L[y'] = pL[y] - y(0)$$

$$L[y''] = p^2 L[y] - py(0) - y'(0)$$

Example 7: Use Table 5 to find the Laplace transform of $g(x) = x^3e^{5x}$.

The presence of the factor e^{5x} suggests using the shifting formula with $k = 5$. Since

$$L[f(x)] = L[x^3] = \frac{3!}{p^{3+1}} = \frac{6}{p^4} = F(p)$$

the shifting formula says that the Laplace transform of $f(x)e^{5x} = x^3e^{5x}$ is equal to $F(p - 5)$. In other words, the Laplace transform of x^3e^{5x} is equal to the Laplace transform of x^3 with the argument p shifted to $p - 5$:

$$L[x^3e^{5x}] = \frac{6}{(p - 5)^4} \quad ■$$

Example 8: Use Table 5 to find the Laplace transform of $f(x) = e^{-2x} \sin x - 3$.

First, since $L[\sin x] = 1/(p^2 + 1)$, the shifting formula (with $k = -2$) says

$$L[e^{-2x} \sin x] = \frac{1}{(p + 2)^2 + 1}$$

Now, because $L[3] = 3 \cdot L[1] = 3/p$, linearity implies

$$L[e^{-2x} \sin x - 3] = \frac{1}{(p + 2)^2 + 1} - \frac{3}{p} \quad ■$$

Example 9: Use Table 5 to find a continuous function whose Laplace transform is $F(p) = 12/p^5$.

This example introduces the idea of the **inverse Laplace transform operator,** L^{-1}. The operator L^{-1} will "un-do" the action of L. Symbolically,

$$f(x) \xrightarrow{L} F(p)$$

$$f(x) \xleftarrow[L^{-1}]{} F(p)$$

If you think of the operator L as changing $f(x)$ into $F(p)$, then the operator L^{-1} just changes $F(p)$ back into $f(x)$. Like L, the inverse operator L^{-1} is linear.

More formally, the result of applying L^{-1} to a function $F(p)$ is to recover the continuous function $f(x)$ whose Laplace transform is the given $F(p)$. [This situation should remind you of the operators D and I (which are, basically, inverses of one another). Each will un-do the action of the other in the sense that if, say, I changes $f(x)$ into $F(x)$, then D will change $F(x)$ back into $f(x)$. In other words, $D = I^{-1}$, so if you apply I and then D, you're back where you started.]

Using Table 5 (reading it from right to left),

$$L^{-1}\left[\frac{12}{p^5}\right] = L^{-1}\left[\frac{\frac{1}{2}\cdot 24}{p^5}\right] = \tfrac{1}{2}L^{-1}\left[\frac{24}{p^5}\right] = \tfrac{1}{2}L^{-1}\left[\frac{4!}{p^{4+1}}\right] = \tfrac{1}{2}x^4 \qquad \blacksquare$$

Example 10: Find the continuous function whose Laplace transform is $F(p) = 1/(p^2 - 1)$.

By partial fraction decomposition,

$$\frac{1}{p^2 - 1} = \frac{-\frac{1}{2}}{p + 1} + \frac{\frac{1}{2}}{p - 1}$$

Therefore, by linearity of L^{-1},

$$L^{-1}\left[\frac{1}{p^2 - 1}\right] = L^{-1}\left[\frac{-\frac{1}{2}}{p + 1}\right] + L^{-1}\left[\frac{\frac{1}{2}}{p - 1}\right] = -\tfrac{1}{2}e^{-x} + \tfrac{1}{2}e^{x}$$

$$= \tfrac{1}{2}(e^x - e^{-x}) \quad \blacksquare$$

Example 11: Determine $L^{-1}\left[\dfrac{12}{(p + 2)^2 + 9}\right]$.

First, note that p has been shifted to $p + 2 = p - (-2)$. Therefore, since

$$L^{-1}\left[\frac{12}{p^2 + 9}\right] = L^{-1}\left[\frac{4 \cdot 3}{p^2 + 9}\right] = 4 \cdot L^{-1}\left[\frac{3}{p^2 + 3^2}\right] = 4 \sin 3x$$

the shifting formula (with $k = -2$) implies

$$L^{-1}\left[\frac{12}{(p + 2)^2 + 9}\right] = 4e^{-2x} \sin 3x \quad \blacksquare$$

Example 12: Evaluate $L^{-1}\left[\dfrac{p - 1}{p^2 - 6p + 25}\right]$.

Although $p^2 - 6p + 25$ cannot be factored over the integers, it can be expressed as the sum of two squares:

$$p^2 - 6p + 25 = (p^2 - 6p + 9) + 16 = (p - 3)^2 + 4^2$$

Therefore,

$$
\begin{aligned}
L^{-1}\left[\frac{p-1}{p^2-6p+25}\right] &= L^{-1}\left[\frac{p-1}{(p-3)^2+4^2}\right] \\
&= L^{-1}\left[\frac{(p-3)+2}{(p-3)^2+4^2}\right] \\
&= L^{-1}\left[\frac{p-3}{(p-3)^2+4^2}\right] + L^{-1}\left[\frac{2}{(p-3)^2+4^2}\right] \\
&= L^{-1}\left[\frac{p-3}{(p-3)^2+4^2}\right] + \tfrac{1}{2}L^{-1}\left[\frac{4}{(p-3)^2+4^2}\right] \\
&= e^{3x}\cos 4x + \tfrac{1}{2}e^{3x}\sin 4x \\
&= e^{3x}(\cos 4x + \tfrac{1}{2}\sin 4x) \quad \blacksquare
\end{aligned}
$$

Using the Laplace Transform to Solve Differential Equations

In this section, you will learn how to use the Laplace transform operator to solve (first- and second-order) differential equations with constant coefficients. In particular, the differential equations must be IVP's with the initial condition(s) specified at $x = 0$.

The method is simple to describe. Given an IVP, apply the Laplace transform operator to both sides of the differential equation. This will transform the differential equation into an *algebraic* equation whose unknown, $F(p)$, is the Laplace transform of the desired solution. Once you solve this algebraic equation for $F(p)$, take the inverse Laplace transform of both sides; the result is the solution to the original IVP.

Before this process is undertaken, it is necessary to see what the Laplace transform operator does to y' and y''. Integration by parts yields

$$\begin{aligned} L[y'] &= \int_0^\infty e^{-px} y'(x)\,dx \\ &= e^{-px} y(x)\Big]_0^\infty - \int_0^\infty -pe^{-px} y(x)\,dx \\ &= -y(0) + p\int_0^\infty e^{-px} y(x)\,dx \qquad (p > 0) \\ &= -y(0) + pL[y] \end{aligned}$$

so

$$L[y'] = pL[y] - y(0)$$

Replacing y by y' in this result gives the Laplace transform of y'':

$$\begin{aligned} L[y''] &= pL[y'] - y'(0) \\ &= p[pL[y] - y(0)] - y'(0) \\ L[y''] &= p^2L[y] - py(0) - y'(0) \end{aligned}$$

Example 13: Use the Laplace transform operator to solve the IVP

$$y' - 2y = e^{3x}$$
$$y(0) = -5$$

Apply the operator L to both sides of the differential equation; then use linearity, the initial condition, and Table 5 to solve for $L[y]$:

$$L[y' - 2y] = L[e^{3x}]$$

$$L[y'] - 2L[y] = L[e^{3x}]$$

$$pL[y] - y(0) - 2L[y] = L[e^{3x}]$$

$$pL[y] + 5 - 2L[y] = \frac{1}{p-3}$$

$$(p-2)L[y] = \frac{1}{p-3} - 5$$

$$L[y] = \frac{-5p + 16}{(p-2)(p-3)}$$

Therefore,

$$y = L^{-1}\left[\frac{-5p + 16}{(p-2)(p-3)}\right]$$

By partial fraction decomposition,

$$\frac{-5p + 16}{(p-2)(p-3)} = \frac{-6}{p-2} + \frac{1}{p-3}$$

so

$$y = L^{-1}\left[\frac{-6}{p-2} + \frac{1}{p-3}\right] = -6e^{2x} + e^{3x}$$

is the solution of the IVP. ■

Usually when faced with an IVP, you *first* find the general solution of the differential equation and *then* use the initial condition(s) to evaluate the constant(s). By contrast, the Laplace transform method uses the initial conditions at the beginning of the solution so that the result obtained in the final step by taking the inverse Laplace transform automatically has the constants evaluated.

Example 14: Use Laplace transforms to solve

$$y'' - 3y' - 4y = -16x$$
$$y(0) = -4$$
$$y'(0) = -5$$

Apply the operator L to both sides of the differential equation; then use linearity, the initial conditions, and Table 5 to solve for $L[y]$:

$$L[y'' - 3y' - 4y] = L[-16x]$$

$$L[y''] - 3L[y'] - 4L[y] = L[-16x]$$

$$[p^2L[y] - py(0) - y'(0)] - 3[pL[y] - y(0)] - 4L[y] = L[-16x]$$

$$[p^2L[y] + 4p + 5] - 3[pL[y] + 4] - 4L[y] = -\frac{16}{p^2}$$

$$(p^2 - 3p - 4)L[y] + 4p - 7 = -\frac{16}{p^2}$$

$$L[y] = \frac{-\frac{16}{p^2} - 4p + 7}{p^2 - 3p - 4}$$

But the partial fraction decomposition of this expression for $L[y]$ is

$$\frac{-\frac{16}{p^2} - 4p + 7}{p^2 - 3p - 4} = \frac{-\frac{16}{p^2} - 4p + 7}{p^2 - 3p - 4} \cdot \frac{p^2}{p^2} = \frac{-16 - 4p^3 + 7p^2}{p^2(p + 1)(p - 4)}$$

$$= \frac{-3}{p} + \frac{4}{p^2} + \frac{1}{p + 1} + \frac{-2}{p - 4}$$

Therefore,

$$y = L^{-1}\left[\frac{-3}{p} + \frac{4}{p^2} + \frac{1}{p+1} + \frac{-2}{p-4}\right]$$

which yields

$$y = -3 + 4x + e^{-x} - 2e^{4x} \qquad \blacksquare$$

Example 15: Use Laplace transforms to determine the solution of the IVP

$$\begin{aligned} y'' - 2y' + 5y &= 0 \\ y(0) &= -1 \\ y'(0) &= 7 \end{aligned}$$

Apply the operator L to both sides of the differential equation; then use linearity, the initial conditions, and Table 5 to solve for $L[y]$:

$$\begin{aligned} L[y'' - 2y' + 5y] &= L[0] \\ L[y''] - 2L[y'] + 5L[y] &= 0 \\ [p^2L[y] - py(0) - y'(0)] - 2[pL[y] - y(0)] + 5L[y] &= 0 \\ [p^2L[y] + p - 7] - 2[pL[y] + 1] + 5L[y] &= 0 \\ (p^2 - 2p + 5)L[y] + p - 9 &= 0 \\ L[y] &= \frac{-p + 9}{p^2 - 2p + 5} \end{aligned}$$

Now,

$$\frac{-p+9}{p^2-2p+5} = \frac{-p+9}{(p^2-2p+1)+4}$$

$$= \frac{-p+9}{(p-1)^2+4}$$

$$= \frac{-(p-1)+8}{(p-1)^2+2^2}$$

$$= \frac{-(p-1)}{(p-1)^2+2^2} + \frac{4\cdot 2}{(p-1)^2+2^2}$$

so

$$y = L^{-1}\left[\frac{-(p-1)}{(p-1)^2+2^2}\right] + 4\cdot L^{-1}\left[\frac{2}{(p-1)^2+2^2}\right]$$

$$= -e^x \cos 2x + 4e^x \sin 2x$$

or more simply,

$$y = e^x\,(4 \sin 2x - \cos 2x) \quad \blacksquare$$

Example 16: Use the fact that if $f(x) = L^{-1}[F(p)]$, then for any positive constant k,

$$L^{-1}[e^{-kp}F(p)] = \begin{cases} 0 & \text{for } x < k \\ f(x-k) & \text{for } x \geq k \end{cases}$$

to solve and sketch the solution of the IVP

$$y' - y = \sigma(x)$$

$$y(0) = 0$$

where σ is the step function

$$\sigma(x) = \begin{cases} 0 & \text{for } 0 \le x < 2 \\ 1 & \text{for } x \ge 2 \end{cases}$$

shown in Figure 9:

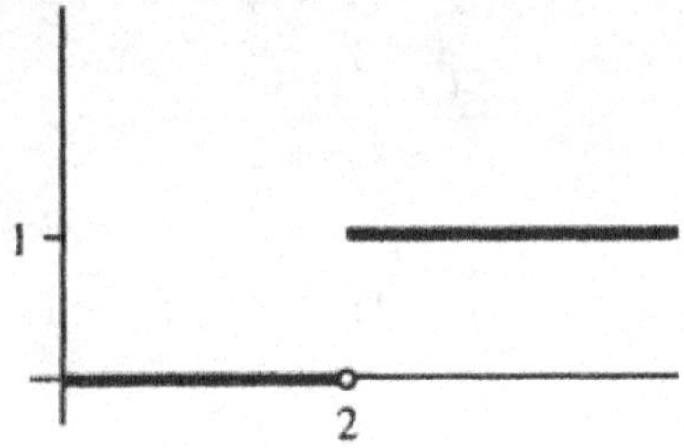

■ Figure 9 ■

As mentioned earlier, the Laplace transform method is particularly well-suited to solving IVP's that involve discontinuous functions such as the step function σ shown above.

As usual, begin by taking the Laplace transform of both sides of the differential equation:

$$L[y' - y] = L[\sigma(x)] \qquad (*)$$

Since $y(0) = 0$, the left-hand side of (*) reduces to

$$L[y' - y] = L[y'] - L[y] = [pL[y] - y(0)] - L[y] = (p - 1)L[y]$$

Using the definition of L, the right-hand side of (*) is now evaluated:

$$L[\sigma(x)] = \int_0^\infty e^{-px}\sigma(x)\,dx = \int_2^\infty e^{-px}\,dx = -\frac{1}{p}e^{-px}\Big]_2^\infty = \frac{e^{-2p}}{p}$$

Therefore, the transformed equation (*) reads

$$(p-1)L[y] = \frac{e^{-2p}}{p}$$

so

$$y = L^{-1}\left[\frac{e^{-2p}}{p(p-1)}\right]$$

But

$$\frac{e^{-2p}}{p(p-1)} = e^{-2p}\frac{1}{p(p-1)} = e^{-2p}\left[\frac{-1}{p} + \frac{1}{p-1}\right]$$

so

$$y = L^{-1}\left[e^{-2p}\frac{1}{p-1}\right] - L^{-1}\left[e^{-2p}\frac{1}{p}\right] \qquad (**)$$

Now, since $L^{-1}[1/(p-1)] = e^x$, the formula given in the statement of the problem says

$$L^{-1}\left[e^{-2p}\frac{1}{p-1}\right] = \begin{cases} 0 & \text{for } x < 2 \\ e^{x-2} & \text{for } x \geq 2 \end{cases}$$

and since $L^{-1}[1/p] \equiv 1$, applying the formula given in the statement of the problem again yields

$$L^{-1}\left[e^{-2p}\frac{1}{p}\right] = \begin{cases} 0 & \text{for } x < 2 \\ 1 & \text{for } x \geq 2 \end{cases}$$

Alternatively, simply notice that

$$L[\sigma(x)] = \frac{e^{-2p}}{p} \Rightarrow L^{-1}\left[e^{-2p}\frac{1}{p}\right] = \sigma(x) = \begin{cases} 0 & \text{for } x < 2 \\ 1 & \text{for } x \geq 2 \end{cases}$$

Substituting these results into (**) gives the solution of the IVP:

$$y = \begin{cases} 0 & \text{for } x < 2 \\ e^{x-2} & \text{for } x \geq 2 \end{cases} - \begin{cases} 0 & \text{for } x < 2 \\ 1 & \text{for } x \geq 2 \end{cases}$$

which becomes

$$y = \begin{cases} 0 & \text{for } x < 2 \\ e^{x-2} - 1 & \text{for } x \geq 2 \end{cases}$$

This function is sketched in Figure 10:

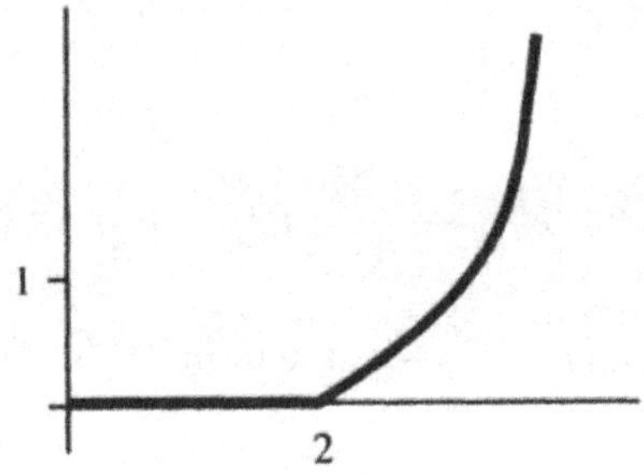

■ Figure 10 ■

It can be argued that the greatest importance of studying differential equations lies in their applications to science. Since the process of predicting and describing the changes in a physical system is perhaps the raison d'être of science, and because the rates at which changes occur are expressed by derivatives, it is not surprising that so many scientific laws are formulated as differential equations. It is not possible here to list the tremendous number of practical applications of differential equations, but this book would be incomplete if there weren't some acknowledgment of their role outside pure mathematics.

Applications of First-Order Equations

Orthogonal trajectories. The term **orthogonal** means *perpendicular,* and **trajectory** means *path* or *curve.* **Orthogonal trajectories,** therefore, are two families of curves that always intersect perpendicularly. A pair of intersecting curves will be perpendicular if the product of their slopes is -1, that is, if the slope of one is the negative reciprocal of the slope of the other. Since the slope of a curve is given by the derivative, two familes of curves $f_1(x, y, c) = 0$ and $f_2(x, y, c) = 0$ (where c is a parameter) will be orthogonal wherever they intersect if

$$\frac{df_2}{dx} = \frac{-1}{df_1/dx}$$

Example 1: The electrostatic field created by a positive point charge is pictured as a collection of straight lines which radiate away from the charge (Figure 11). Using the fact that the *equipotentials* (surfaces of constant electric potential) are orthogonal to the electric field lines, determine the geometry of the equipotentials of a point charge.

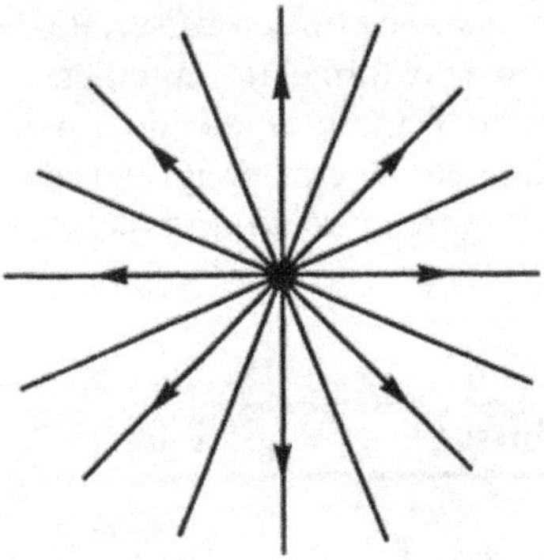

■ Figure 11 ■

If the origin of an xy coordinate system is placed at the charge, then the electric field lines can be described by the family

$$y = cx$$

The first step in determining the orthogonal trajectories is to obtain an expression for the slope of the curves in this family that does *not* involve the parameter c. In the present case,

$$y = cx \Rightarrow \frac{dy}{dx} = c = \frac{y}{x} \qquad (*)$$

The differential equation describing the orthogonal trajectories is therefore

$$\frac{dy}{dx} = -\frac{x}{y} \qquad (**)$$

since the right-hand side of (**) is the negative reciprocal of the right-hand side of (*). Because this equation is separable, the solution can proceed as follows:

$$\frac{dy}{dx} = -\frac{x}{y}$$

$$y\,dy = -x\,dx$$

$$\int y\,dy = \int -x\,dx$$

$$\tfrac{1}{2}y^2 = -\tfrac{1}{2}x^2 + c'$$

$$x^2 + y^2 = c^2$$

where $c^2 = 2c'$.

The equipotential lines (that is, the intersection of the equipotential surfaces with any plane containing the charge) are therefore the family of circles $x^2 + y^2 = c^2$ centered at the origin. The equipotential and electric field lines for a point charge are shown in Figure 12.

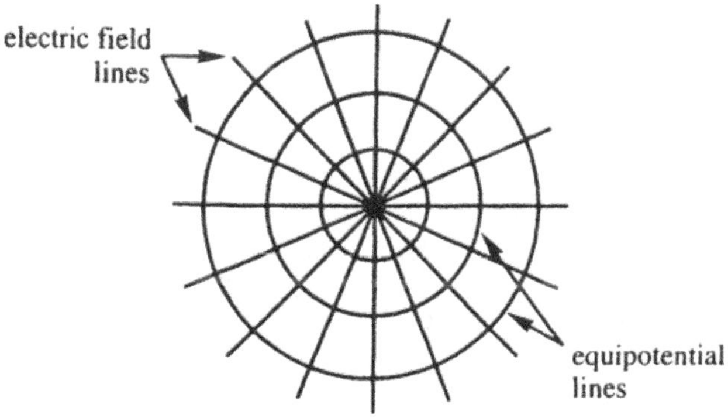

■ Figure 12 ■

Example 2: Determine the orthogonal trajectories of the family of circles $x^2 + (y - c)^2 = c^2$ tangent to the x axis at the origin.

The first step is to determine an expression for the slope of the curves in this family that does not involve the parameter c. By implicit differentiation,

$$x^2 + (y - c)^2 = c^2$$

$$x^2 + y^2 = 2cy$$

$$2x + 2y\frac{dy}{dx} = 2c\frac{dy}{dx}$$

$$\frac{dy}{dx} = \frac{x}{c - y}$$

To eliminate c, note that

$$x^2 + (y - c)^2 = c^2 \Rightarrow x^2 + y^2 = 2cy \Rightarrow c = \frac{x^2 + y^2}{2y}$$

The expression above for dy/dx may now be written in the form

$$\frac{dy}{dx} = \frac{x}{c - y} = \frac{x}{\dfrac{x^2 + y^2}{2y} - y} = \frac{2xy}{x^2 - y^2} \qquad (*)$$

Therefore, the differential equation describing the orthogonal trajectories is

$$\frac{dy}{dx} = \frac{y^2 - x^2}{2xy} \qquad (**)$$

since the right-hand side of (**) is the negative reciprocal of the right-hand side of (*).

If equation (**) is written in the form

$$(y^2 - x^2)\,dx - 2xy\,dy = 0$$

note that it is not exact (since $M_y = 2y$ but $N_x = -2y$). However, because

$$\frac{M_y - N_x}{N} = \frac{2y - (-2y)}{-2xy} = -\frac{2}{x}$$

is a function of x alone, the differential equation has

$$\mu = e^{\int(-2/x)dx} = e^{-2\ln x} = e^{\ln(x^{-2})} = x^{-2}$$

as an integrating factor. After multiplying through by $\mu = x^{-2}$, the differential equation describing the desired family of orthogonal trajectories becomes

$$(x^{-2}y^2 - 1)\,dx - 2x^{-1}y\,dy = 0$$

which is now exact (because $M_y = 2x^{-2}y = N_x$). Since

$$\int \overline{M}\,\partial x = \int (x^{-2}y^2 - 1)\,\partial x = -x^{-1}y^2 - x$$

and

$$\int \overline{N}\,\partial y = \int (-2x^{-1}y)\,\partial y = -x^{-1}y^2$$

the solution of the differential equation is

$$-x^{-1}y^2 - x = -2c$$

(The reason the constant was written as $-2c$ rather than as c will be apparent in the calculation below.) With a little algebra, the equation for this family may be rewritten:

$$-x^{-1}y^2 - x = -2c$$

$$x^{-1}y^2 + x = 2c$$

$$y^2 + x^2 = 2cx$$

$$(x - c)^2 + y^2 = c^2$$

This shows that the orthogonal trajectories of the circles tangent to the x axis at the origin are the circles tangent to the y axis at the origin! See Figure 13.

■ Figure 13 ■

Radioactive decay. Some nuclei are energetically unstable and can spontaneously transform into more stable forms by various processes known collectively as **radioactive decay.** The rate at which a particular radioactive sample will decay depends on the identity of the sample. Tables have been compiled which list the half-lives of various radioisotopes. The **half-life** is the amount of time required for one-half the nuclei in a sample of the isotope to decay; therefore, the shorter the half-life, the more rapid the decay rate.

The rate at which a sample decays is proportional to the amount of the sample present. Therefore, if $x(t)$ denotes the amount of a radioactive substance present at time t, then

$$\frac{dx}{dt} = -kx \qquad (k > 0)$$

(The rate dx/dt is negative, since x is decreasing.) The positive constant k is called the **rate constant** for the particular radioisotope. The solution of this separable first-order equation is

$$x = x_0 e^{-kt} \qquad (*)$$

where x_0 denotes the amount of substance present at time $t = 0$. The graph of this equation (Figure 14) is known as the **exponential decay curve:**

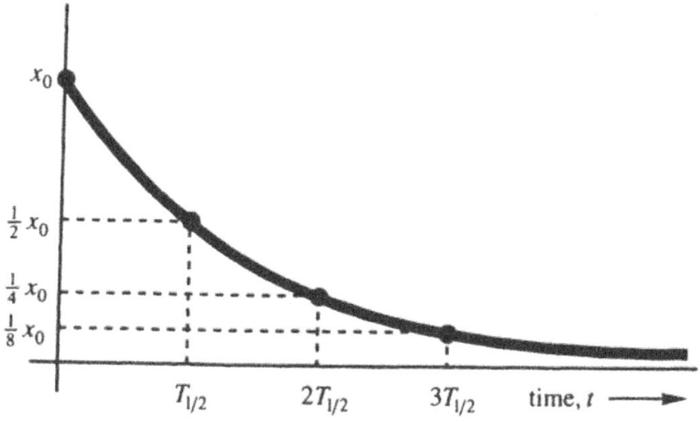

Figure 14

The relationship between the half-life (denoted $T_{1/2}$) and the rate constant k can easily be found. Since, by definition, $x = \frac{1}{2}x_0$ at $t = T_{1/2}$, (*) becomes

$$\tfrac{1}{2}x_0 = x_0 e^{-kT_{1/2}} \Rightarrow -kT_{1/2} = \ln \tfrac{1}{2} \Rightarrow k = \frac{\ln 2}{T_{1/2}}$$

Because the half-life and rate constant are inversely proportional, the shorter the half-life, the greater the rate constant, and, consequently, the more rapid the decay.

Radiocarbon dating is a process used by anthropologists and archaeologists to estimate the age of organic matter (such as wood or bone). The vast majority of carbon on earth is nonradioactive carbon-12 (^{12}C). However, cosmic rays cause the formation of **carbon-14** (^{14}C), a radioactive isotope of carbon which becomes incorporated into living plants (and therefore into animals) through the intake of radioactive carbon dioxide ($^{14}CO_2$). When the plant or animal dies, it ceases its intake of carbon-14, and the amount present at the time of death begins to decrease (since the ^{14}C decays and is not replenished). Since the half-life of ^{14}C is known to be 5730 years, by measuring the concentration of ^{14}C in a sample, its age can be determined.

Example 3: A fragment of bone is discovered to contain 20% of the usual ^{14}C concentration. Estimate the age of the bone.

The relative amount of ^{14}C in the bone has decreased to 20% of its original value (that is, the value when the animal was alive). Thus, the problem is to calculate the value of t at which $x(t) = 0.20x_0$ (where x = the amount of ^{14}C present). Since

$$k = \frac{\ln 2}{T_{1/2}} = \frac{\ln 2}{5730}$$

the exponential decay equation (*) says

$$0.20x_0 = x_0e^{-[(\ln 2)/5730]t}$$

$$\ln (0.20) = -\frac{\ln 2}{5730}t$$

$$t = 5730\frac{-\ln (0.20)}{\ln 2}$$

$$t \approx 13{,}300 \text{ years} \quad \blacksquare$$

Newton's Law of Cooling. When a hot object is placed in a cool room, the object dissipates heat to the surroundings, and its temperature decreases. **Newton's Law of Cooling** states that the rate at which the object's temperature decreases is proportional to the difference between the temperature of the object and the ambient temperature. At the beginning of the cooling process, the difference between these temperatures is greatest, so this is when the rate of temperature decrease is greatest. However, as the object cools, the temperature difference gets smaller, and the cooling rate decreases; thus, the object cools more and more slowly as time passes. To formulate this process mathematically, let $T(t)$ denote the temperature of the object at time t and let T_s denote the (essentially constant) temperature of the surroundings. Newton's Law of Cooling then says

$$\frac{dT}{dt} = -k(T - T_s) \qquad (k > 0)$$

Since $T_s < T$ (that is, since the room is cooler than the object), T decreases, so the rate of change of its temperature, dT/dt, is necessarily negative. The solution of this separable differential equation proceeds as follows:

$$\begin{aligned} \int \frac{dT}{T - T_s} &= \int -k\,dt \\ \ln(T - T_s) &= -kt + c' \\ T - T_s &= e^{-kt+c'} \\ T &= T_s + ce^{-kt} \qquad (*) \end{aligned}$$

Example 4: A cup of coffee (temperature = 190°F) is placed in a room whose temperature is 70°F. After five minutes, the temperature of the coffee has dropped to 160°F. How many more minutes must elapse before the temperature of the coffee is 130°F?

Assuming that the coffee obeys Newton's Law of Cooling, its temperature T as a function of time is given by equation (*) with $T_s = 70$:

$$T(t) = 70 + ce^{-kt}$$

Because $T(0) = 190$, the value of the constant of integration (c) can be evaluated:

$$T(0) = 190 \Rightarrow 190 = 70 + c \Rightarrow c = 120$$

Furthermore, since information about the cooling rate is provided ($T = 160$ at time $t = 5$ minutes), the cooling constant k can be determined:

$$160 = T(5)$$

$$160 = 70 + 120e^{-5k}$$

$$e^{-5k} = \tfrac{3}{4}$$

$$-5k = \ln \tfrac{3}{4}$$

$$k = \tfrac{1}{5} \ln \tfrac{4}{3}$$

Therefore, the temperature of the coffee t minutes after it is placed in the room is

$$T(t) = 70 + 120e^{-[(1/5)\ln(4/3)]t}$$

Now, setting $T = 130$ and solving for t yields

$$130 = 70 + 120e^{-[(1/5)\ln(4/3)]t}$$

$$e^{-[(1/5)\ln(4/3)]t} = \tfrac{1}{2}$$

$$(-\tfrac{1}{5}\ln\tfrac{4}{3})t = \ln\tfrac{1}{2}$$

$$t = \frac{5\ln 2}{\ln\frac{4}{3}}$$

$$t \approx 12 \text{ minutes}$$

This is the *total* amount of time after the coffee is initially placed in the room for its temperature to drop to 130°F. Therefore, after waiting five minutes for the coffee to cool from 190°F to 160°F, it is necessary to then wait an additional seven minutes for it to cool down to 130°F. ■

Skydiving (part I). Once a sky diver jumps from an airplane, there are two forces that determine her motion: the pull of the earth's gravity and the opposing force of air resistance. At high speeds, the strength of the air resistance force (the *drag force*) can be expressed as kv^2, where v is the speed with which the sky diver descends and k is a proportionality constant determined by such factors as the diver's cross-sectional area and the viscosity of the air. Once the parachute opens, the descent speed decreases greatly, and the strength of the air resistance force is given by Kv.

Newton's Second Law states that if a net force F_{net} acts on an object of mass m, the object will experience an acceleration a given by the simple equation

$$F_{net} = ma$$

Since the acceleration is the time derivative of the velocity, this law can be expressed in the form

$$F_{\text{net}} = m\frac{dv}{dt} \qquad (*)$$

In the case of a sky diver initially falling without a parachute, the drag force is $F_{\text{drag}} = kv^2$, and the equation of motion (*) becomes

$$mg - kv^2 = m\frac{dv}{dt}$$

or more simply,

$$\frac{dv}{dt} = g - bv^2$$

where $b = k/m$. [The letter g denotes the value of the *gravitational acceleration,* and mg is the force due to gravity acting on the mass m (that is, mg is its weight). Near the surface of the earth, g is approximately 9.8 meters per second^2.] Once the sky diver's descent speed reaches $v = \sqrt{g/b} = \sqrt{mg/k}$, the equation above says $dv/dt = 0$; that is, v stays constant. This occurs when the speed is great enough for the force of air resistance to balance the weight of the sky diver; the net force and (consequently) the acceleration drop to zero. This constant descent velocity is known as the **terminal velocity.** For a sky diver falling in the spread-eagle position without a parachute, the value of the proportionality constant k in the drag equation $F_{\text{drag}} = kv^2$ is approximately $\frac{1}{4}$ kg/m. Therefore, if the sky diver has a total mass of 70 kg (which corresponds to a weight of about 150 pounds), her terminal velocity is

$$\underset{\text{(no parachute)}}{v_{\text{terminal}}} = \sqrt{\frac{mg}{k}} = \sqrt{\frac{(70)(9.8)}{\frac{1}{4}}} \approx 52 \text{ m/s}$$

or approximately 120 miles per hour.

Once the parachute opens, the air resistance force becomes $F_{\text{air resist}} = Kv$, and the equation of motion (*) becomes

$$mg - Kv = m\frac{dv}{dt}$$

or more simply,

$$\frac{dv}{dt} = g - Bv$$

where $B = K/m$. Once the parachutist's descent speed slows to $v = g/B = mg/K$, the equation above says $dv/dt = 0$; that is, v stays constant. This occurs when the speed is low enough for the weight of the sky diver to balance the force of air resistance; the net force and (consequently) the acceleration reach zero. Again, this constant descent velocity is known as the *terminal velocity.* For a sky diver falling *with* a parachute, the value of the proportionality constant K in the equation $F_{\text{air resist}} = Kv$ is approximately 110 kg/s. Therefore, if the sky diver has a total mass of 70 kg, the terminal velocity (with the parachute open) is only

$$v_{\substack{\text{terminal}\\ \text{(with parachute)}}} = \frac{mg}{K} = \frac{(70)(9.8)}{110} \approx 6.2 \text{ m/s}$$

which is about 14 miles per hour. Since it is safer to hit the ground while falling at a rate of 14 miles per hour rather than at 120 miles per hour, sky divers use parachutes.

Example 5: After a free-falling sky diver of mass m reaches a constant velocity of v_1, her parachute opens, and the resulting air resistance force has strength Kv. Derive an equation for the speed of the sky diver t seconds after the parachute opens.

As stated above, once the parachute opens, the equation of motion is

$$\frac{dv}{dt} = g - Bv$$

where $B = K/m$. The parameter that will arise from the solution of this first-order differential equation will be determined by the initial condition $v(0) = v_1$ (since the sky diver's velocity is v_1 at the moment the parachute opens, and the "clock" is reset to $t = 0$ at this instant). This separable equation is solved as follows:

$$\int \frac{dv}{g - Bv} = \int dt$$

$$-\frac{1}{B}\ln(g - Bv) = t + c''$$

$$\ln(g - Bv) = -Bt + c'$$

$$g - Bv = ce^{-Bt}$$

Now, since $v(0) = v_1 \Rightarrow g - Bv_1 = c$, the desired equation for the sky diver's speed t seconds after the parachute opens is

$$g - Bv = (g - Bv_1)e^{-Bt}$$

$$Bv = g - (g - Bv_1)e^{-Bt}$$

$$v = \frac{g}{B}\left[1 - \left(1 - \frac{B}{g}v_1\right)e^{-Bt}\right]$$

$$v = \frac{mg}{K}\left[1 + \left(\frac{K}{mg}v_1 - 1\right)e^{-(K/m)t}\right]$$

Note that as time passes (that is, as t increases), the term $e^{-(K/m)t}$ goes to zero, so (as expected) the parachutist's speed v slows to mg/K, which is the terminal speed with the parachute open. ■

Applications of Second-Order Equations

Skydiving (part II). The principal quantities used to describe the motion of an object are position (s), velocity (v), and acceleration (a). Since velocity is the time derivative of the position, and acceleration is the time derivative of the velocity, acceleration is the second time derivative of the position. Therefore, the position function $s(t)$ for a moving object can be determined by writing Newton's Second Law, $F_{net} = ma$, in the form

$$F_{net} = m\frac{d^2s}{dt^2}$$

and solving this second-order differential equation for s.

[You may see the derivative with respect to time represented by a *dot*. For example, $\dot{s}$ ("s dot") denotes the first derivative of s with respect to t, and $\ddot{s}$ ("s double dot") denotes the second derivative of s with respect to t. The dot notation is used only for derivatives with respect to *time*.]

Example 6: A sky diver (mass m) falls long enough without a parachute (so the drag force has strength kv^2) to reach her first terminal velocity (denoted v_1). When her parachute opens, the air resistance force has strength Kv. At what minimum altitude must her parachute open so that she slows to within 1% of her new (much lower) terminal velocity (v_2) by the time she hits the ground?

Let y denote the vertical distance measured downward from the point at which her parachute opens (which will be designated time $t = 0$). Then Newton's Second Law ($F_{\text{net}} = ma$) becomes $mg - Kv = ma$, or, since $v = \dot{y}$ and $a = \ddot{y}$,

$$mg - K\dot{y} = m\ddot{y}$$

This situation is therefore described by the IVP

$$\ddot{y} + \frac{K}{m}\dot{y} = g$$

$$y(0) = 0$$

$$\dot{y}(0) = v_1$$

The differential equation is second-order linear with constant coefficients, and its corresponding homogeneous equation is

$$\ddot{y} + B\dot{y} = 0$$

where $B = K/m$. The auxiliary polynomial equation, $r^2 + Br = 0$, has $r = 0$ and $r = -B$ as roots. Since these are real and distinct, the general solution of the corresponding homogeneous equation is

$$y_h = c_1 + c_2e^{-Bt} = c_1 + c_2e^{-(K/m)t}$$

The given nonhomogeneous equation has $\bar{y} = (mg/K)t$ as a particular solution, so its general solution is

$$y = c_1 + c_2e^{-(K/m)t} + \frac{mg}{K}t \qquad (*)$$

Now, to apply the initial conditions and evaluate the parameters c_1 and c_2:

$$y(0) = 0 \Rightarrow \left[c_1 + c_2 e^{-(K/m)t} + \frac{mg}{K} t \right]_{t=0} = 0 \Rightarrow c_1 + c_2 = 0$$

$$\dot{y}(0) = v_1 \Rightarrow \left[-c_2 \frac{K}{m} e^{-(K/m)t} + \frac{mg}{K} \right]_{t=0} = v_1 \Rightarrow -c_2 \frac{K}{m} + \frac{mg}{K} = v_1$$

These two equations imply

$$c_1 = \frac{m}{K}\left(v_1 - \frac{mg}{K}\right) \quad \text{and} \quad c_2 = -\frac{m}{K}\left(v_1 - \frac{mg}{K}\right)$$

Once these values are substituted into (*), the complete solution to the IVP can be written as

$$y = \frac{mg}{K}\left[t + \left(\frac{v_1}{g} - \frac{m}{K}\right)\left(1 - e^{-(K/m)t}\right) \right] \qquad (**)$$

The derivative of this expression gives the velocity of the sky diver t seconds after the parachute opens:

$$v = \dot{y} = \frac{mg}{K}\left[1 + \left(\frac{v_1 K}{mg} - 1\right) e^{-(K/m)t} \right] \qquad (***)$$

Note that equation (***) is equivalent to the result of Example 5 (as it should be).

The question asks for the minimum altitude at which the sky diver's parachute must be open in order to land at a velocity of $(1.01)v_2$. Therefore, set v equal to $(1.01)v_2$ in equation (***) and solve for t; then substitute the result into (**) to find the desired altitude. Omitting the messy details, once the expression in (***) is set equal to $(1.01)v_2$, the value of t is found to be

$$t = \frac{m}{K} \ln \frac{\dfrac{v_1 K}{mg} - 1}{\dfrac{(1.01)v_2 K}{mg} - 1} = \frac{m}{K} \ln \frac{\dfrac{v_1 K}{mg} - 1}{0.01}$$

and substituting this result into (**) yields

$$y = \frac{mg}{K}\left[\frac{m}{K} \ln \frac{\dfrac{v_1 K}{mg} - 1}{0.01} + \left(\frac{v_1}{g} - \frac{m}{K}\right)\left(1 - \frac{0.01}{\dfrac{v_1 K}{mg} - 1}\right)\right]$$

To evaluate the numerical answer, the following values—from the discussion in *Skydiving (part I)*—are used:

mass of sky diver: $m = 70$ kg

gravitational acceleration: $g = 9.8\ \text{m/s}^2$

air resistance proportionality constant: $K = 110$ kg/s

terminal velocity without parachute: $v_1 = \sqrt{\dfrac{mg}{k}} \approx 52$ m/s

terminal velocity with parachute: $v_2 = \dfrac{mg}{K} \approx 6.2$ m/s

These substitutions give a descent time t [the time interval between the parachute opening to the point where a speed of $(1.01)v_2$ is attained] of approximately 4.2 seconds, and a minimum altitude at which the parachute must be opened of $y \approx 55$ meters (a little higher than 180 feet). ■

Simple harmonic motion. Consider a spring fastened to a wall, with a block attached to its free end at rest on an essentially frictionless horizontal table. The block can be set into motion by pulling or pushing it from its original position and then letting go, or by striking it (that is, by giving the block a nonzero initial velocity). The force exerted by the spring keeps the block oscillating on the tabletop. This is the prototypical example of **simple harmonic motion.**

The force exerted by a spring is given by *Hooke's Law;* this states that if a spring is stretched or compressed a distance x from its natural length, then it exerts a force given by the equation

$$F = -kx$$

The positive constant k is known as the *spring constant* and is directly related to the spring's stiffness: The stiffer the spring, the larger the value of k. The minus sign implies that when the spring is stretched (so that x is positive), the spring pulls back (because F is negative), and conversely, when the spring is compressed (so that x is negative), the spring pushes outward (because F is positive). Therefore, the spring is said to exert a *restoring force,* since it always tries to restore the block to its *equilibrium* position (the position where the spring is neither stretched nor compressed). The restoring force here is proportional to the displacement ($F = -kx \propto x$), and it is for this reason that the resulting **periodic** (regularly repeating) motion is called *simple harmonic.*

Newton's Second Law can be applied to this spring-block system. Once the block is set into motion, the only horizontal force that acts on it is the restoring force of the spring. Therefore, the equation $F_{\text{net}} = ma$ becomes $-kx = m\ddot{x}$, or

$$\ddot{x} + \frac{k}{m}x = 0$$

This is a homogeneous second-order linear equation with constant coefficients. The auxiliary polynomial equation is $r^2 + \frac{k}{m} = 0$, which has distinct conjugate complex roots $\pm i\sqrt{k/m}$. Therefore, the general solution of this differential equation is

$$x = c_1 \cos \sqrt{\frac{k}{m}}\,t + c_2 \sin \sqrt{\frac{k}{m}}\,t \qquad (*)$$

This expression gives the displacement of the block from its equilibrium position (which is designated $x = 0$).

Example 7: A block of mass 1 kg is attached to a spring with force constant $k = \frac{25}{4}$ N/m. It is pulled $\frac{3}{10}$ m from its equilibrium position and released from rest. Obtain an equation for its position at any time t; then determine how long it takes the block to complete one cycle (one round trip).

All that is required is to adapt equation (*) to the present situation. First, since the block is released from rest, its intial velocity is 0:

$$\dot{x}(0) = 0 \Rightarrow \left[-c_1 \sqrt{\frac{k}{m}} \sin \sqrt{\frac{k}{m}}\, t + c_2 \sqrt{\frac{k}{m}} \cos \sqrt{\frac{k}{m}}\, t \right]_{t=0} = 0$$

$$\Rightarrow c_2 = 0$$

Since $c_2 = 0$, equation (*) reduces to $x = c_1 \cos \sqrt{k/m}\, t$. Now, since $x(0) = +\frac{3}{10}$ m, the remaining parameter can be evaluated:

$$x(0) = \frac{3}{10} \Rightarrow \left[c_1 \cos \sqrt{\frac{k}{m}}\, t \right]_{t=0} = \frac{3}{10} \Rightarrow c_1 = \frac{3}{10}$$

Finally, since $k = \frac{25}{4}$ N/m and $m = 1$ kg, $\sqrt{k/m} = \sqrt{25/4} = \frac{5}{2}$. Therefore, the equation for the position of the block as a function of time is given by

$$x = \tfrac{3}{10} \cos \tfrac{5}{2} t$$

where x is measured in meters from the equilibrium position of the block. This function is *periodic,* which means it repeats itself at regular intervals. The cosine and sine functions each have a period of 2π, which means every time the argument increases by 2π, the function returns to its previous value. (Recall that if, say, $x = \cos \theta$, then θ is called the **argument** of the cosine function.) The argument here is $\frac{5}{4}t$, and $\frac{5}{4}t$ will increase by 2π every time t increases by $\frac{4}{5}\pi$. Therefore, this block will complete one cycle, that is, return to its original position ($x = \frac{3}{10}$ m), every $\frac{4}{5}\pi \approx 2.5$ seconds. ■

The length of time required to complete one cycle (one round trip) is called the **period** of the motion (and denoted by T). It can be shown in general that for the spring-block oscillator,

$$T = 2\pi \sqrt{\frac{m}{k}}$$

Note that the period does not depend on where the block started, only on its mass and the stiffness of the spring. The maximum distance (greatest displacement) from equilibrium is called the **amplitude** of the motion. Therefore, it makes no difference whether the block oscillates with an amplitude of 2 cm or 10 cm; the period will be the same in either case. This is one of the defining characteristics of simple harmonic motion: the period is independent of the amplitude.

Another important characteristic of an oscillator is the number of cycles that can be completed per unit time; this is called the **frequency** of the motion [denoted traditionally by ν (the Greek letter nu) but less confusingly by the letter f]. Since the period specifies the length of time per cycle, the number of cycles per unit time (the frequency) is simply the reciprocal of the period: $f = 1/T$. Therefore, for the spring-block simple harmonic oscillator,

$$f = \frac{1}{2\pi} \sqrt{\frac{k}{m}}$$

Frequency is usually expressed in *hertz* (abbreviated Hz); 1 Hz equals 1 cycle per second.

The quantity $\sqrt{k/m}$ (the coefficient of t in the argument of the sine and cosine in the general solution of the differential equation describing simple harmonic motion) appears so often in problems of this type that it is given its own name and symbol. It is called the **angular frequency** of the motion and denoted by ω (the Greek letter omega). Note that $\omega = 2\pi f$.

Damped oscillations. The spring-block oscillator discussed above is an idealized example of a frictionless system. In real life, however, frictional (or *dissipative*) forces must be taken into account, particularly if you want to model the behavior of the system over a long period of time. Unless the block slides back and forth on a frictionless table in a room evacuated of air, there will be resistance to the block's motion due to the air (just as there is for a falling sky diver). This resistance would be rather small, however, so you may want to picture the spring-block apparatus submerged in a large container of clear oil. The viscosity of the oil will have a profound effect upon the block's oscillations. The air (or oil) provides a **damping force,** which is proportional to the velocity of the object. (Again, recall the sky diver falling with a parachute. At the relatively low speeds attained with an open parachute, the force due to air resistance was given as Kv, which is proportional to the velocity.)

With a restoring force given by $-kx$ and a damping force given by $-Kv$ (the minus sign means that the damping force opposes the velocity), Newton's Second Law ($F_{\text{net}} = ma$) becomes $-kx - Kv = ma$, or, since $v = \dot{x}$ and $a = \ddot{x}$,

$$-kx - K\dot{x} = m\ddot{x}$$

This second-order linear differential equation with constant coefficients can be expressed in the more standard form

$$m\ddot{x} + K\dot{x} + kx = 0$$

The auxiliary polynomial equation is $mr^2 + Kr + k = 0$, whose roots are

$$r = \frac{-K \pm \sqrt{K^2 - 4mk}}{2m}$$

The system will exhibit periodic motion only if these roots are distinct conjugate complex numbers, because only then will the general solution of the differential equation involve the periodic functions sine and cosine. In order for this to be the case, the discriminant $K^2 - 4mk$ must be negative; that is, the damping constant K must be small; specifically, it must be less than $2\sqrt{mk}$. When this happens, the motion is said to be **underdamped,** because the damping is not so great as to prevent the system from oscillating; it just causes the amplitude of the oscillations to gradually die out. [If the damping constant K is too great, then the discriminant is nonnegative, the roots of the auxiliary polynomial equation are real (and negative), and the general solution of the differential equation involves only decaying exponentials. This implies there would be no sustained oscillations.]

In the underdamped case ($K < 2\sqrt{mk}$), the roots of the auxiliary polynomial equation can be written as

$$r = -\frac{K}{2m} \pm i\,\frac{\sqrt{4mk - K^2}}{2m} = -\frac{K}{2m} \pm i\sqrt{\frac{k}{m} - \frac{K^2}{4m^2}}$$

and consequently, the general solution of the defining differential equation is

$$x = e^{-(K/2m)t}\left(c_1 \cos\sqrt{\frac{k}{m} - \frac{K^2}{4m^2}}\,t + c_2 \sin\sqrt{\frac{k}{m} - \frac{K^2}{4m^2}}\,t\right)$$

Example 8: (Compare to Example 7.) A block of mass 1 kg is attached to a spring with force constant $k = \frac{25}{4}$ N/m. It is pulled $\frac{3}{10}$ m from its equilibrium position and released from rest. If this spring-block apparatus is submerged in a viscous fluid medium which exerts a damping force of $-4v$ (where v is the instantaneous velocity of the block), sketch the curve that describes the position of the block as a function of time.

The net force on the block is $-\frac{25}{4}x - 4v = -\frac{25}{4}x - 4\dot{x}$, so Newton's Second Law becomes

$$\ddot{x} + 4\dot{x} + \frac{25}{4}x = 0$$

because $m = 1$. Since the roots of the auxiliary polynomial equation, $r^2 + 4r + \frac{25}{4} = 0$, are

$$r = \frac{-4 \pm \sqrt{4^2 - 4 \cdot 1 \cdot \frac{25}{4}}}{2 \cdot 1} = \frac{-4 \pm 3i}{2} = -2 \pm \frac{3}{2}i$$

the general solution of the differential equation is

$$x = e^{-2t}(c_1 \cos \tfrac{3}{2}t + c_2 \sin \tfrac{3}{2}t)$$

Because the block is released from rest, $v(0) = \dot{x}(0) = 0$:

$$[e^{-2t}(-\tfrac{3}{2}c_1 \sin \tfrac{3}{2}t + \tfrac{3}{2}c_2 \cos \tfrac{3}{2}t) - 2e^{-2t}(c_1 \cos \tfrac{3}{2}t + c_2 \sin \tfrac{3}{2}t)]_{t=0} = 0$$

This implies $\frac{3}{2}c_2 - 2c_1 = 0$. And, since $x(0) = \frac{3}{10}$ m,

$$[e^{-2t}(c_1 \cos \tfrac{3}{2}t + c_2 \sin \tfrac{3}{2}t)]_{t=0} = \tfrac{3}{10} \Rightarrow c_1 = \tfrac{3}{10}$$

Therefore, $c_2 = \frac{2}{3}(2c_1) = \frac{4}{3}(\frac{3}{10}) = \frac{4}{10}$, and the equation that gives the position of the block as a function of time is

$$x = \tfrac{1}{10}e^{-2t}(3 \cos \tfrac{3}{2}t + 4 \sin \tfrac{3}{2}t)$$

where x is measured in meters from the equilibrium position of the block.

This expression for the position function can be rewritten using the trigonometric identity $\cos(\alpha - \beta) = \cos\alpha\cos\beta + \sin\alpha\sin\beta$, as follows:

$$x = \tfrac{1}{10}e^{-2t}(3\cos\tfrac{3}{2}t + 4\sin\tfrac{3}{2}t)$$

$$= \tfrac{1}{2}e^{-2t}[(\cos\tfrac{3}{2}t)(\tfrac{3}{5}) + (\sin\tfrac{3}{2}t)(\tfrac{4}{5})]$$

$$= \tfrac{1}{2}e^{-2t}[(\cos\tfrac{3}{2}t)(\cos\phi) + (\sin\tfrac{3}{2}t)(\sin\phi)]$$

$$x = \tfrac{1}{2}e^{-2t}\cos(\tfrac{3}{2}t - \phi)$$

The *phase angle*, ϕ, is defined here by the equations $\cos\phi = \frac{3}{5}$ and $\sin\phi = \frac{4}{5}$, or, more briefly, as the first-quadrant angle whose tangent is $\frac{4}{3}$ (it's the larger acute angle in a 3-4-5 right triangle). The presence of the decaying exponential factor e^{-2t} in the equation for $x(t)$ means that as time passes (that is, as t increases), the amplitude of the oscillations gradually dies out. See Figure 15.

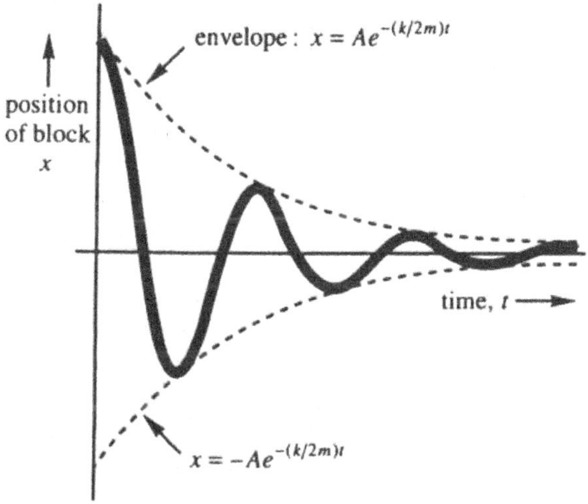

■ Figure 15 ■

The angular frequency of this periodic motion is the coefficient of t in the cosine, $\omega' = \frac{3}{2}$ rad/s, which implies a period of

$$T' = \frac{1}{f'} = \frac{2\pi}{\omega'} = \frac{2\pi}{\frac{3}{2}} = \frac{4}{3}\pi \approx 4.2 \text{ seconds}$$

Compare this to Example 7, which described the same spring, block, and initial conditions but with no damping. The position function there was $x = \frac{3}{10} \cos \frac{5}{2}t$; it had constant amplitude, an angular frequency of $\omega = \frac{5}{2}$ rad/s, and a period of just $\frac{4}{5}\pi \approx 2.5$ seconds. Therefore, not only does (under)damping cause the amplitude to gradually die out, but it also increases the period of the motion. But this seems reasonable: Damping reduces the speed of the block, so it takes longer to complete a round trip (hence the increase in the period). This will always happen in the case of underdamping, since $\omega' = \sqrt{k/m - K^2/4m^2}$ will always be lower than $\omega = \sqrt{k/m}$. ■

Electric circuits and resonance. When an electric circuit containing an ac voltage source, an inductor, a capacitor, and a resistor in series is analyzed mathematically, the equation that results is a second-order linear differential equation with constant coefficients. The voltage $v(t)$ produced by the ac source will be expressed by the equation $v = V \sin \omega t$, where V is the maximum voltage generated. An *inductor* is a circuit element that opposes changes in current, causing a voltage drop of $L(di/dt)$, where i is the instantaneous current and L is a proportionality constant known as the *inductance*. A *capacitor* stores charge, and when each plate carries a magnitude of charge q, the voltage drop across the capacitor is q/C, where C is a constant called the *capacitance*. Finally, a *resistor* opposes the flow of current, creating a voltage drop equal to iR, where the constant R is the *resistance*. *Kirchhoff's Loop Rule* states that the algebraic sum of the voltage differences as one goes around any closed loop in a circuit is equal to zero. Therefore, if the voltage source, inductor, capacitor, and resistor are all in series, then

$$V\sin\omega t - L\frac{di}{dt} - \frac{q}{C} - iR = 0$$

which can be rewritten as

$$L\frac{di}{dt} + Ri + \frac{1}{C}q = V\sin\omega t$$

Now, if an expression for $i(t)$—the current in the circuit as a function of time—is desired, then the equation to be solved must be written in terms of i. To this end, differentiate the equation directly above, and use the definition $i \equiv dq/dt$:

$$L\frac{d^2i}{dt^2} + R\frac{di}{dt} + \frac{1}{C}i = \omega V\cos\omega t \qquad (*)$$

This differential equation governs the behavior of an **LRC series circuit** with a source of sinusoidally varying voltage.

The first step in solving this equation is to obtain the general solution of the corresponding homogeneous equation

$$L\frac{d^2i}{dt^2} + R\frac{di}{dt} + \frac{1}{C}i = 0 \qquad (**)$$

But notice that this differential equation has exactly the same mathematical form as the equation for the damped oscillator,

$$m\frac{d^2x}{dt^2} + K\frac{dx}{dt} + kx = 0 \qquad (***)$$

By comparing the two equations, it is easy to see that the current (i) is analogous to the position (x), the inductance (L) is analogous to the mass (m), the resistance (R) is analogous to the damping constant (K), and the reciprocal capacitance ($1/C$) is analogous to the spring constant (k). Since the general solution of (***) was found to be

$$x = e^{-(K/2m)t}\left(c_1 \cos \sqrt{\frac{k}{m} - \frac{K^2}{4m^2}}\,t + c_2 \sin \sqrt{\frac{k}{m} - \frac{K^2}{4m^2}}\,t\right)$$

$$\text{if } K < 2\sqrt{mk}$$

the general solution of (**) must be, by analogy,

$$i_h = e^{-(R/2L)t}\left(c_1 \cos \sqrt{\frac{1}{LC} - \frac{R^2}{4L^2}}\,t + c_2 \sin \sqrt{\frac{1}{LC} - \frac{R^2}{4L^2}}\,t\right)$$

$$\text{if } R < 2\sqrt{L/C}$$

But the solution does not end here. The original differential equation (*) for the LRC circuit was nonhomogeneous, so a particular solution must still be obtained. The family of the nonhomogeneous right-hand term, $\omega V \cos \omega t$, is $\{\sin \omega t, \cos \omega t\}$, so a particular solution will have the form

$$\bar{i} = A \sin \omega t + B \cos \omega t$$

where A and B are the undetermined coefficients. Given this expression for $\bar{i}$, it is easy to calculate

$$\bar{i}' = \omega A \cos \omega t - \omega B \sin \omega t$$

$$\bar{i}'' = -\omega^2 A \sin \omega t - \omega^2 B \cos \omega t$$

Substituting these last three expressions into the given nonhomogeneous differential equation (*) yields

$$\left[-\left(\omega^2 L - \frac{1}{C}\right)A - \omega RB\right] \sin \omega t$$

$$+ \left[(\omega R)A - \left(\omega^2 L - \frac{1}{C}\right)B\right] \cos \omega t = \omega V \cos \omega t$$

Therefore, in order for this to be an identity, A and B must satisfy the simultaneous equations

$$-\left(\omega^2 L - \frac{1}{C}\right)A - (\omega R)B = 0$$

$$(\omega R)A - \left(\omega^2 L - \frac{1}{C}\right)B = \omega V$$

The solution of this system is

$$A = \frac{\omega^2 RV}{\left(\omega^2 L - \frac{1}{C}\right)^2 + (\omega R)^2} \quad \text{and} \quad B = \frac{-\omega V\left(\omega^2 L - \frac{1}{C}\right)}{\left(\omega^2 L - \frac{1}{C}\right)^2 + (\omega R)^2}$$

These expressions can be simplified by invoking the following standard definitions:

- ωL is called the *inductive reactance* and denoted X_L.
- $\frac{1}{\omega C}$ is called the *capacitive reactance* and denoted X_C
- $X_L - X_C$ is simply called the *reactance* and denoted X
- $\sqrt{(X_L - X_C)^2 + R^2} = \sqrt{X^2 + R^2}$ is called the *impedance* and denoted Z

Therefore,

$$\begin{aligned}\left(\omega^2 L - \frac{1}{C}\right)^2 + (\omega R)^2 &= \left[\omega\left(\omega L - \frac{1}{\omega C}\right)\right]^2 + (\omega R)^2 \\ &= \omega^2\left[\left(\omega L - \frac{1}{\omega C}\right)^2 + R^2\right] \\ &= \omega^2[(X_L - X_C)^2 + R^2] \\ &= \omega^2[X^2 + R^2] \\ &= \omega^2 Z^2\end{aligned}$$

and the expressions for the coefficients A and B given above can be written as

$$A = \frac{\omega^2 RV}{\left(\omega^2 L - \frac{1}{C}\right)^2 + (\omega R)^2} = \frac{\omega^2 RV}{\omega^2 Z^2} = \frac{RV}{Z^2}$$

$$B = \frac{-\omega V\left(\omega^2 L - \frac{1}{C}\right)}{\left(\omega^2 L - \frac{1}{C}\right)^2 + (\omega R)^2} = \frac{-\omega V \cdot \omega\left(\omega L - \frac{1}{\omega C}\right)}{\omega^2 Z^2}$$

$$= \frac{-\omega^2 V(X_L - X_C)}{\omega^2 Z^2} = -\frac{VX}{Z^2}$$

These simplifications yield the following particular solution of the given nonhomogeneous differential equation:

$$\bar{i} = \frac{RV}{Z^2}\sin \omega t - \frac{VX}{Z^2}\cos \omega t = \frac{V}{Z}\left(\frac{R}{Z}\sin \omega t - \frac{X}{Z}\cos \omega t\right)$$

Combining this with the general solution of the corresponding homogeneous equation gives the complete solution of the nonhomogeneous equation: $i = i_h + \bar{i}$ or

$$i = e^{-(R/2L)t}\left(c_1 \cos \sqrt{\frac{1}{LC} - \frac{R^2}{4L^2}}\,t + c_2 \sin \sqrt{\frac{1}{LC} - \frac{R^2}{4L^2}}\,t\right) + \frac{V}{Z}\left(\frac{R}{Z}\sin \omega t - \frac{X}{Z}\cos \omega t\right)$$

Despite its rather formidable appearance, it lends itself easily to analysis. The first term [the one with the exponential-decay factor $e^{-(R/2L)t}$] goes to zero as t increases, while the second term remains indefinitely. For these reasons, the first term is known as the *transient current*, and the second is called the *steady-state current*:

$$i = \underbrace{e^{-(R/2L)t}\left(c_1 \cos\sqrt{\frac{1}{LC} - \frac{R^2}{4L^2}}\,t + c_2 \sin\sqrt{\frac{1}{LC} - \frac{R^2}{4L^2}}\,t\right)}_{\substack{\text{transient current}\\ \text{(goes to 0)}}}$$

$$+ \underbrace{\frac{V}{Z}\left(\frac{R}{Z}\sin \omega t - \frac{X}{Z}\cos \omega t\right)}_{\text{steady-state current}}$$

Example 9: Consider the underdamped LRC series circuit discussed above. Once the transient current becomes so small that it may be neglected, under what conditions will the amplitude of the oscillating steady-state current be maximized? In particular, assuming that the inductance L, capacitance C, resistance R, and voltage amplitude V are fixed, how should the angular frequency ω of the voltage source be adjusted to maximize the steady-state current in the circuit?

The steady-state curent is given by the equation

$$i = \frac{V}{Z}\left(\frac{R}{Z}\sin \omega t - \frac{X}{Z}\cos \omega t\right)$$

By analogy with the phase-angle calculation in Example 8, this equation is rewritten as follows:

$$\begin{aligned} i &= \frac{V}{Z}\left(\frac{R}{Z}\sin \omega t - \frac{X}{Z}\cos \omega t\right) \\ &= \frac{V}{Z}\left(\frac{R}{\sqrt{X^2 + R^2}}\sin \omega t - \frac{X}{\sqrt{X^2 + R^2}}\cos \omega t\right) \\ &= \frac{V}{Z}(\cos \phi \sin \omega t - \sin \phi \cos \omega t) \\ &= \frac{V}{Z}\sin(\omega t - \phi) \end{aligned}$$

(where $\cos\phi = R/\sqrt{X^2 + R^2}$ and $\sin\phi = X/\sqrt{X^2 + R^2}$). Therefore, the amplitude of the steady-state current is V/Z, and, since V is fixed, the way to maximize V/Z is to minimize Z. Because $Z = \sqrt{X^2 + R^2}$, Z will be minimized if $X = 0$. And because ω is necessarily positive,

$$X = 0 \Leftrightarrow X_L - X_C = 0 \Leftrightarrow \omega L = \frac{1}{\omega C} \Leftrightarrow \omega^2 = \frac{1}{LC} \Leftrightarrow \omega = \frac{1}{\sqrt{LC}}$$

This value of ω is called the **resonant angular frequency.** When the underdamped circuit is "tuned" to this value, the steady-state current is maximized, and the circuit is said to be **in resonance.** This is the principle behind tuning a radio, the process of obtaining the strongest response to a particular transmission. In this case, the frequency (and therefore angular frequency) of the transmission is fixed (an FM station may be broadcasting at a frequency of, say, 95.5 MHz, which actually means that it's broadcasting in a narrow *band* around 95.5 MHz), and the value of the capacitance C or inductance L can be varied by turning a dial or pushing a button. According to the calculation above, resonance is achieved when

$$\omega = \frac{1}{\sqrt{LC}}$$

Therefore, in terms of a (relatively) fixed ω and a variable capacitance, resonance will occur when

$$C = \frac{1}{\omega^2 L} = \frac{1}{(2\pi f)^2 L}$$

(where f is the frequency of the broadcast). Or in terms of a variable inductance, the circuitry will resonate to a particular station when L is adjusted to the value

$$L = \frac{1}{\omega^2 C} = \frac{1}{(2\pi f)^2 C} \qquad \blacksquare$$

Made in the USA
Monee, IL
11 November 2025

34421606R00108